Extraterrestrial Communication Code

The Discovery, Meaning, and Our Response to Their Message

Stephen J. Silva

Stephen J. Silva has over thirty years of professional experience as a civil engineer and engineering project manager. He holds a Civil/Environmental Engineering degree from the University of Vermont, as well as a degree in Marine Technology from the Florida Institute of Technology.
Email: etcommgroup@gmail.com
Blog: etcommgroup.com

Disclaimer
All pictures, diagrams, and other images in this book that are not originals created by the author were taken from "public domain" locations. There is no intent to include proprietary images or words in this book.

Copyright ©2021 by Stephen Silva

No part of this publication may be reproduced, distributed, or transmitted in any form or by any means, including photocopying, recording, or other electronic or mechanical methods, or by any information storage and retrieval system, without the prior written permission of the publisher, except in the case of very brief quotations embodied in critical reviews, and certain other noncommercial uses permitted by copyright law.

Book design by eBook DesignWorks

This book is dedicated to all of my wonderful family
with whom I have been blessed.

It is also dedicated to all those who understand
that there is intelligent life in the universe beyond Earth
and are committed to establishing communications
with that intelligent life for the benefit of humankind.

Contents

Introduction 1

1.
The Concept 3

2.
Back to Basics 6

3.
The Search 20

4.
The Model 34

5.
Model Results 45

6.
Analysis of Result Words 50

7.
Response to the Message 87

8.
Response Implementation 102

9.
Proposed Experiment 107

Introduction

THIS BOOK IS ABOUT THE discovery of a way to establish communication with extraterrestrials (ETs) that has never before been conceived. It was developed by discovering and interpreting indicators that may have been intentionally provided by ETs for us to find and figure out.

When Apollo 17 took the "blue marble" picture of Earth from outer space in 1972, it made the world pause and consider that the Earth has limits and that we are all one human race living in a fragile bubble. It was a short pause. Sadly, human rights abuses continue, as does our adverse impact on our environment.

Perhaps the only way we will come to understand and respect our fragility is by establishing communication with intelligent life from a place not of this Earth. Only then will we truly understand that we are tiny in the universe, we are one human race, and that we need to come together as one people of Earth in order to enter the community of intelligent life in the universe. Communicating with extraterrestrials will focus our attention in that direction and bring the world together as one.

> "The human animal cannot be trusted for anything good except en masse. The combined thought and action of the whole people of any race, creed or nationality will always point in the right direction."
>
> —*Harry S. Truman*

1.

The Concept

THERE IS A LOT OF evidence in the form of ancient megastructures and texts, photographs, videos, and eyewitness accounts to suggest that ETs have historically had an influence on the development of humanity as we know it today. Many people believe that ETs still come and go from Earth all the time. The one thing—and the most important thing—that is missing in this unsolved mystery is communication with ETs. We have tried to reach out to the stars and send messages, but as far as we know, there has been no response. That being the case, one has to wonder why they don't respond to our messages. They seem to keep coming and going all the time, so what are we missing?

Of all the things to worry about, I often fall asleep at night thinking that there must be a better way to establish communication with ETs—a better and simpler way than what has been tried in the past, and what is being attempted now. These thoughts are always about finding the "conduit" for communicating with ETs, and the premise that the conduit must be reasonably simple and logical. Establishing communication must surely be simpler than the messages we have already launched to the stars in hopes of a response that has apparently failed so far. What can we conceive of or do that is different?

One evening, I drifted off to sleep with that exact question burning in my head. Seemingly out of nowhere, simple numerical patterns appeared to me in some sort of dream or vision. I know what a dream is, and this was different. I woke up, remembered the patterns, and wrote them down. I then wrote a series of steps or a procedure to follow using those patterns. It popped into my head in vivid color and detail. At that moment, the exact and entire path was unclear, but the first

> Seemingly out of nowhere, simple numerical patterns appeared to me in some sort of dream or vision.

steps down that path were very clear. Days later, I looked at what I had jotted down and felt the need to take the first steps. I just had to take my vision and test the possibilities of what it might mean. The code and its meaning that resulted from that test is described here. I have used the process to build a mathematical model, decoded the message given from the model's output, and interpreted what that message might mean and how to respond. The process correlates some very interesting factual information from the fields of science, history, mathematics, geography, and logic. It also has just the right amount of deductive reasoning sprinkled into the mix. It all flows together, leading to an amazing conclusion, and forming the major concept behind this book.

As a civil engineer by trade, my professional life has largely been about finding solutions to the problems of civilization, including the management of natural resources and protecting the natural environment for the betterment of humanity. I have also always had an interest in the subject of intelligent life in the universe beyond the boundaries of Earth. As I followed my vision and researched the subject, I found that it raised more questions than answers. Over time, these questions became so alluring that it became necessary to conduct deeper independent research in addition to reading about the theories posed by others. What I discovered is the key to unlocking a message that was probably developed and then hidden by ETs for humans to eventually figure out. It is a message much like the messages we send out into the universe in hopes of an ET response; however, the difference lies in how the message has been delivered, revealed, and interpreted. Make no mistake: I tested many ideas and theories that led to dead ends, forcing me to stop and rethink all of the existing evidence and mainstream theories, including my own. I concluded that there must be something that scientists are missing in the quest for ET communication and in terms of information about how they travel across the universe, and their historic and ongoing presence on this Earth.

By going back to basics, some of the answers are unveiled. When you find and use the keys, it is relatively simple to unlock the door; what lies on the other side of the door is not so simple. The door opens into a vast new area of exploration and discovery. If you think about it, that is exactly what ETs would have wanted to provide—a relatively simple communication code that reveals much once you've figured out the first step. To get there, we need to rethink everything and go back to basics.

2.
Back to Basics

THE SEARCH FOR INTELLIGENT LIFE in the universe has become incredibly complex (at least for Earthlings) with respect to the coded messages we launch into the universe and the technology we use to launch those messages. There is conflicting information on the subject coming from credible sources who believe ETs exist, and from those who do not. There are a lot of well-educated, professional, and reputable people who claim they have had direct ET contact already, but they can't prove it with physical evidence. Governments "officially" claim they have never communicated with an ET, but there seems to be a high probability that there is physical evidence to the contrary, and it is being kept hidden from the public. Humanity, in general, has a tendency to believe ETs exist, and that is why we continue to search and reach out to the stars. However, despite all of our efforts, we have not yet established "communication" as far as the general public is aware. There is a difference between "contact" and "communication." There seems to be overwhelming documented evidence that ETs have traveled to Earth throughout history and still visit Earth to this day. The evidence varies from descriptions and drawings in ancient texts, carvings on ancient monuments and structures, and eyewitness accounts. The US government still denies this is evidence at all. In all cases, the contact or evidence of contact was initiated by the ETs and not by us, and that fact is important to the premise of this book. That fact also seems to be the common thread that runs throughout all of the documentation. We need to take a step back and reboot our thinking… Perhaps we don't need to spend so much time searching for proof of ET existence because we already have that figured out. Instead, we should be focusing our efforts on establishing communication with ETs. Perhaps the search has become unfocused—too complicated and off-track. In this book, I take a step back and think of the problem differently. A back-to-basics view and a build-up-from-there approach, if you will.

The first basic concept to accept is that intelligent life exists in many places in the universe. The only differences between each of the forms of intelligent life are: 1) the level of advancement based on how long that life has been cognizant, and 2) the environment in which that life developed and evolved over the history of the universe. Life everywhere in the universe shares some fundamental needs, such as food, water, and shelter; the capacity for curiosity, however, is a character trait reserved for more intelligent life. Satisfying our curiosity about life beyond our world is one of the character traits we obviously share with ETs, who have reached beyond the boundaries of their planet to travel to this tiny, blue marble in space we call Earth. This is not to say that there is no intelligent life on Earth other than humans. For all we know, dolphins and whales and other big-brained Earthly creatures already communicate with ETs. It has been documented that the radio frequencies of whale song are identical to the frequencies of signals picked up by some radio telescopes listening for transmissions from outer space. That, however, is a subject for another day.

Exploration and discovery have been part of basic human nature for as long as we have had the courage to stand tall and take those first tentative steps toward the horizon. We are always looking over the next hill just to see what might be there.

Is the desire to explore the nature of intelligent life unique to humans? Or is it part of the nature of intelligent life everywhere in the universe? If we consider all that we think we know about ETs, the answer must be that intelligent life everywhere in the universe is driven to explore the unknown. That is how intelligent life develops: it reaches out to find answers to what is unknown.

The subject of ETs having a presence on this Earth and having an influence on humanity has fascinated people for eons, from ancient civilizations to this day. As

a result, there is a lot of readily available information out there to research. The topic is as old as humanity and can get very complicated very quickly. Through all the twists and turns in the documentation, it seems to all boil down to two basic things. The first is trying to understand the universe and how it works. The second is understanding ETs in the universe. By the most basic definition, an ET is a "being" that originates and exists outside the Earth or its atmosphere. If you think about it, we are all ETs to somebody somewhere out there in the universe.

Technically, ET status includes all forms of life, either complex and intelligent, as well as much less developed, i.e., single-celled organisms. For the purposes of this book, an ET, unless otherwise specified, will only refer to intelligent life capable of intentionally traveling to Earth as many believe they have been doing for hundreds or even thousands of years. With that boundary set, it is reasonable to surmise that ETs must be more technologically advanced than humans because they have probably been in existence for a much longer time, giving them longer to advance and develop. Perhaps, however, ETs are not very complex or different from humans with respect to basic instincts and beliefs. If we could reset the clock of human and ET development back to day one, and track our respective development side by side, would we be equally "advanced" after thousands of years? Maybe not exactly equal,

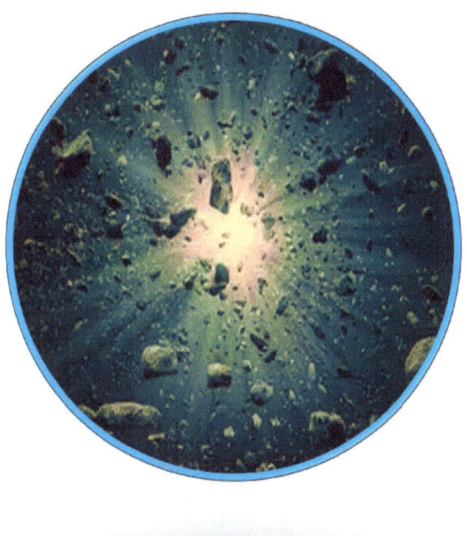

but the differences might not be so great. Do you ever wonder what ETs theorize about the creation of the universe? There are many people who believe that God created the universe and all the creatures within the universe, on whatever planet in whatever galaxy they dwell. If you are not a person of faith, the concept of God doing all that is not even remotely possible. What are the other possibilities that we theorize to explain the universe and its creation? Was it the Big Bang that created the universe? What exactly is the Big Bang theory? We hear the words all the time, but do we really know what they mean?

The author of the theory itself was a Belgian physicist, cosmologist, and Catholic priest named Georges Lemaitre.

Astronomer and mathematician Fred Hoyle is generally credited with inventing the term "Big Bang" in 1949. He used the term on BBC radio during his promotion of panspermia theory as the origin of life on Earth. Panspermia is a hypothesis stating that life exists throughout the universe and that it is distributed by celestial bodies or possibly space vehicles unintentionally carrying microorganisms. Hoyle was a brilliant man who lived from 1915 to 2001, and his ideas deserve further study. Hoyle coined "Big Bang" as something of a dismissive term for Lemaitre's premise that Hoyle thought had no basis.

Ironically, the Big Bang theory has become the most well-known and widely accepted theory for the universe's creation, even though it is generally misunderstood by laypeople who don't know a lot about astrophysics. The main misunderstanding is that the Big Bang theory describes the creation of the universe.

If you dig into it, that description is not completely accurate. The Big Bang theory is not exactly what the average person thinks it is. The Big Bang idea theorizes that the universe developed from a minuscule dense thing that expanded into the indescribably vast universe as we understand it today. However, the universe existed

The Big Bang theory is not exactly what the average person thinks it is

before the Big Bang. The Big Bang does not theorize that some originally minuscule but ultra-dense thing "exploded" to create the universe, nor does it theorize what came before the Big Bang. There is no good explanation of what was beyond the "universe" before it theoretically began. What environment did that tiny, ultra-dense thing that exploded and expanded exist within?

Most people will usually say that they think the Big Bang was the explosion that created the universe; that perception is not accurate. The Big Bang actually describes the expansion of an *existing* universe. The Big Bang theory is really a belief that there was a moment in time when all the forces of the universe came together as a single unified force that expanded an existing universe—the key word being "existing." There are also the theories of the Steady State, the Eternal Inflation, the Oscillating Universe, and others. So, which is harder to believe—God, or the Big Bang? The basic truth of the matter is that we don't know how the universe was created. We may never know how the universe was created or when it was created. Maybe some other concept beyond the word "creation" is needed that has yet to be conceived. The best we may ever be able to do is to study the behavior of the universe and work with those behaviors for the betterment of humanity. Making claims about the creation of the universe and the history of the universe seems like a big stretch when you consider that we don't even know much beyond a reasonable doubt about the history of life on this Earth. More and more discoveries are being made every day that change (or should change) the history books. The biggest discovery yet to be made is how the presence of ETs has influenced the development of life on Earth. That will be the most significant world-changing discovery ever made, and we are on that trail now.

Of all the theories about the creation of the universe, God is the only one that does not require science, physics, or mathematics. The God idea (Christian or otherwise) has no equations—it only requires faith. Most people of faith believe that miracles happen and that they happen for a reason. There is no Einstein-like equation that has been developed to predict the occurrence of a miracle, or explain a miracle after it happens. By definition, a miracle is an unexpected but positive event that cannot be explained by any known natural or scientific laws. A miracle is the work of God or some other divine entity. Conversely and by definition, a theory is an unproven but possible supposition or collection of ideas intended to explain or predict something based on known scientific, natural, or mathematical laws. Let's leave the topic on that note, because for the purposes of this book, it doesn't matter. What does matter is the basic fact that the universe exists, and within it, so does intelligent life. The probability that the only intelligent life in the universe is on Earth is virtually nil; that is the one thing on which we can all generally agree. The universe is just too vast to be devoid of intelligent life other than ourselves. So how do we establish communication with that intelligent life?

There is a good amount of fairly strong evidence to support the conclusion that ETs are regularly coming and going from Earth, or at least doing flybys. There is also a good amount of evidence to suggest that ETs have been here since humanity's earliest days on Earth, when we only had the ability to document their presence using primitive cave wall diagrams. For all we know, ETs were here long before the human species ever existed, and a lot of people believe exactly that. It is also reasonable to surmise that there is much more evidence on this matter that we have yet to discover. The ultimate proof will be some actual communication with ETs for the entire world to see. We have been trying to get hard evidence of direct communication for decades with no results, so how do we get the answers? Something has to change, and maybe that something is our fundamental thinking.

Let's stay with the back-to-basics approach. Perhaps we should simplify our attempts to establish communication with ETs. One of the most basic and effective "human performance tools" used for establishing clear communication between people and enabling the understanding of instructions is called three-way or repeat-back communication.

This concept and procedure came about after some smart person deduced that most accidents in industrial and other settings occur because of unclear or miscommunicated instructions from one person to another. Three-way communication is part of every industrial worker's training, and it works so well that it is also used by the military.

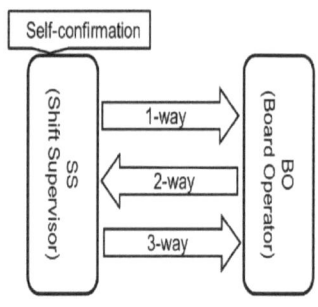

Three-way communication works like this:

1. The sender of the message states the message to the receiving person clearly and concisely. For example, the sender says, *"Bring the pressure up to 100 psi."*

2. The receiver of the message acknowledges the instruction by repeating the message back to the sender. *"Understand that you want me to bring the pressure up to 100 psi."*

3. The sender acknowledges the receiver's response and verbally confirms to the receiver that the message is correct and properly understood. *"Affirmative. Bring the pressure up to 100 psi."* The receiver is now clear to execute the instruction and communicate to the sender that the task is complete.

4. If the sender does not agree with the receiver's reply, the sender must verbally indicate that the two parties do not understand each other, and then the process must start again from the beginning. If the receiver responded by saying, *"Understand that you want me to bring the pressure up to 2,000 psi,"* the sender would respond with: *"Negative. Bring the pressure up to 100 psi."*

5. The process is repeated until the instructions are understood, verified as understood, and the task is completed.

This type of communication process is basic, and it is common in the animal kingdom as well. If you go to a wooded area and listen to the birds, you can often hear this three-way communication in action. One bird sounds out, another bird out of visual contact, but within hearing distance will repeat the noise, and then the sending bird will confirm. It's truly remarkable yet very basic. Ocean mammals such as dolphins and whales do the same thing. Perhaps that is what we need to do to establish communication with ETs.

One could argue that three-way communication only works between like species—birds with the same kind of bird, whales with whales, dolphins with dolphins, etc. That may be somewhat true; however, there are many instances when completely different species have engaged in this three-way process and have succeeded in establishing communication. Take, for example, humans and birds, such as like parrots. "Polly wants a cracker?" The bird has (and needs to have) the ability to repeat back in our language, "Polly wants a cracker." Our response is to give them a cracker—which constitutes food—a basic necessity. The birds are able to respond to our communication in a language that is not theirs, and this results in a reward that meets one of their basic needs, i.e., food.

Now think about trained dolphins, seals, killer whales, etc. What is "trained" to one person is more like being a "prisoner" to another (and certainly must seem so to the animal in the cage, but let's not go there). We provide an instruction; the animal repeats it back (or otherwise demonstrates that they understand our message and the required response), and they get food. Captive animals need us to give them food, so they have no choice but to figure out our communication signals. Perhaps we are not as smart as birds and other animals; perhaps ETs have overestimated our capability to understand their message and respond appropriately. We have not demonstrated that we have understood their message—we don't even know that we've found a message. No cracker for us.

Think about the evolution of human-to-human communication in recent years. We like to think that our communication capability is advancing; however, much of the younger generation seems to be losing their desire or need to communicate directly through spoken language. The decreasing desire of young people to engage in "normal" conversation is becoming increasingly apparent. It is now more common to engage in electronic communication, such as text messages, emails, and social media platforms. Is this what happened with inter-ET communication? Is the future of human-to-human communication moving to electronics and then perhaps to extrasensory perception (ESP)? Who knows? Look around an airport or any public place. Everybody's eyes are glued to their cell phones, and they don't look at each other. Communication tools for humans are changing rapidly, but the basic process is the same: sender sends the message, receiver responds, sender acknowledges. Whatever the species or methodology, the process remains the same.

Many people believe that ET communication has already been established, but that the information is being withheld from the general public by governments. That may or may not be true. What is true is that if we expect the president of the United States to get on national television and announce to the world that we have established communication with ETs, it's probably not going to happen anytime soon. Why would ETs not just make direct contact in an obvious way? Why all the mystery and hiding and searching for codes and signals, etc.? Perhaps it is because we haven't properly executed the three-way communication process. We can do it with animals, after all, so why not with ETs? Perhaps we haven't demonstrated that we've found and understood their first attempt at communication. We need to find their message and repeat it back clearly in order to demonstrate that we share at least minimal likeness to the ETs and are ready for communication.

ETs may not be convinced we are ready for direct contact as equals. We may appear as primitive to them as cavemen seem to us today.

To an advanced ET civilization, what they knew about us in antiquity and what they know about us today may still put us in the "primitive" category in their view. To make an analogy, maybe to ETs we are like the Ewoks from the *Star Wars* movies—nothing more than a reasonably intelligent and aggressive species of two-legged

creature native to the Earth. We are reasonably skilled at survival within our natural environment and at the development of simple technologies, and we seem to be quick learners when exposed to advanced technology. Note that in those movies, communication was developed after basic contact. Some people speculate and defend the theory that, even without direct communication, ETs show up from time to time and leave a crew behind to study our development, observe our "progress," and go about their ET business, mostly out of our sight. Every so often, they intervene in our development to steer us in a certain direction just to see what happens. This could be anything from a DNA "tweak" to providing advanced knowledge and the use of advanced technologies to "move us along."

Establishing ET communication would change our world forever, but maybe it's no big thing to an ET; it will happen when they feel we are ready. Maybe we are just not that interesting to them yet, but they are keeping an eye on us as we develop. When they are satisfied that we are ready to be introduced to the universal community of intelligent life of all types and species, they will communicate with us directly.

Some of the most intelligent scientists in the world are searching for ETs and trying to communicate with them. The SETI Institute, for example, is a nonprofit organization focused on finding intelligent life in the universe beyond Earth. They have close to one hundred scientists, most with PhDs, all on the hunt for

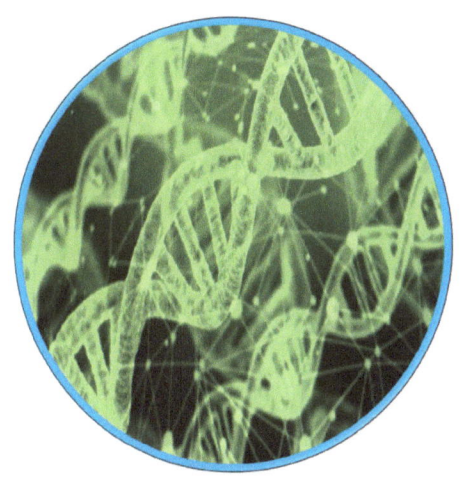

ET. The fact that the search continues at a high level contradicts the theory that communication already happened with ancient cultures because if that was the case, then why are we still searching? There is a difference between "contact" and "communication." All of the information available on the subject still leaves these basic questions unanswered:

1. How did ETs get here?
2. Why were/are they here?
3. How do we communicate?
4. Where do they go in between visits?

Maybe a different approach to answering some of these questions is needed. Every planet is an island to be discovered and explored in the ocean of the universe. What if we think like an ET explorer that found a new island in the universe to explore (Earth)? Whatever the reason for their presence on Earth, the early ETs would have probably left behind some way to let us know they were here. They probably would have left us a way to find them, a way for us to demonstrate that we have advanced enough to find their map and use it to communicate. That is what we humans did when new worlds were discovered here on Earth.

Much like Step 1 in three-way communication, the sender (ETs) states the instructions to the receiver (us). It is now our turn to receive, understand, and "repeat back" the instructions, but we have not done that yet.

The first ETs that came and left probably had no grand plan to return immediately. It was a case of first things first: they probably documented their discovery of Earth, studied us and Earth's resources for a period of time, and then returned home and reported their findings back to their superiors.

This is exactly what we did when we "discovered" new lands and islands on Earth that were previously unknown (to everybody except the natives who had already lived there for hundreds of years). We returned home to report our discoveries, and when we came back to these places, we provided the natives with technologies and tools to help their societies advance. That did not work out so well for the Native Americans, as noted by the late great astrophysicist Stephen Hawking. *"If aliens visit us, the outcome would be much as when Columbus landed in America, which didn't turn out well for the Native Americans. We only have to look at ourselves to see how intelligent life might develop into something we wouldn't want to meet"* (from *Into the Universe with Stephen Hawking*, 2010).

It seems that ETs did or are doing the same sort of thing to the natives of Earth. The basic behaviors and instincts of ETs and humans (discovery and intervention) appear to be similar, as might be expected.

There is solid evidence to suggest that ETs were here on Earth interacting with early humans. They may have been here even before humans came into existence. There are numerous examples of cave drawings that depict what looks like spacecraft and alien life forms. In addition, there are countless ancient text references from all over the world that describe visitors from the sky, many of which also include artworks showing the same. Then there are the giant structures, like the pyramids, that would have been impossible for humans to construct without the use of advanced technology or even divine intervention. There also seems to be an unexplained acceleration in the advancement of humans in comparison with a normal evolutionary timetable. We went from living in caves to constructing complex megalithic structures that align with celestial bodies in the relatively short time of a few thousand years, as far as we can estimate. We went from riding horses to riding rockets to the moon in less than one hundred years. That can't be a natural rate of development.

There are those who believe ETs had something to do with that accelerated rate of development. The timing and concept of the "missing link" seems to be something that needs better explanation and a more accurate definition. It's not just about bones; it's about brains and an accelerated advancement of technology and knowledge in general.

A skeleton was found in Ethiopia in 1974 that was believed for a long time to be the "missing link." The skeleton was given the name Lucy. We have now figured out that Lucy is probably not the missing link. The following article was published by Creation Ministries International in 1990.

"Lucy" isn't the "Missing Link"!

"Lucy" is the popular name given to the famous fossil skeleton found in 1974 in Ethiopia by American anthropologist Donald Johanson. To many people, Lucy is regarded as a certain link between ape-like creatures and man—thus supposedly proving evolution. But is Lucy really a pre-human ancestor?

According to Richard Leakey, who along with Johanson is probably the best-known fossil-anthropologist in the world, Lucy's skull is so incomplete that most of it is "imagination made of plaster of Paris." Leakey even said in 1983 that no firm conclusion could be drawn about what species Lucy belonged to.

In reinforcement of the fact that Lucy is not a creature "in between" ape and man, Dr. Charles Oxnard, Professor of Anatomy and Human Biology at the University of Western Australia, said in 1987 of the australopithecines (the group to which Lucy is said to have belonged):

"The various australopithecines are, indeed, more different from both African apes and humans in most features than these latter are from each other. Part of the basis of this acceptance has been the fact that even opposing investigators have found these large differences as they too used techniques and research designs that were less biased by prior notions as to what the fossils might have been." Oxnard's firm conclusion? "The australopithecines are unique."

Neither Lucy nor any other australopithecine is therefore intermediate between humans and African apes. Nor are they similar enough to humans to be any sort of ancestor of ours.

> *Lucy and the australopithecines show nothing about human evolution, and should not be promoted as having any sort of "missing link" status. The creationist alternative, that humans, apes and other creatures were created that way in the beginning, remains the only explanation consistent with all the evidence.*

The time of the missing link may actually be the time (or one of the many times) when ETs returned and interfered with human development. That is what humans did on Earth. We explored, found "new worlds," returned home to report the discovery, and then returned to the new world and interfered with the development of the native culture. Why would the ET process be any different? After all, in theory, humans and ETs are similar in nature, right? We are just far apart in terms of our developmental timeline.

At some point in time, it seems that ETs decided there was something worthwhile here and then returned to observe and intervene. Is that not exactly what human explorers did after they found a "new world"? Earth explorers who discovered new worlds provided the natives with technology, left indicators of their presence behind, and made maps to show how to get there again and get home. Would we not expect the ETs who first discovered Earth to do the same or similar? Our history books tell us that European explorers discovered the "new world" of the Americas, but it wasn't a new world to the indigenous people they found living there already. The similarity between the actions of European explorers discovering the Americas and those of ETs discovering Earth is something we need to keep in mind. Europeans took America away from the indigenous peoples by force. Are ETs intending to do the same? Will they someday colonize Earth?

3.
The Search

THE SEARCH FOR ET MESSAGES is largely done by listening to radio signals, which we know as radio astronomy, using radio telescopes. Visual media like "star watching" with optical telescopes are more about observing the movements of celestial bodies than they are about ET communication. Radio astronomy is a relatively young branch of "astro-science" that has produced some of the most important discoveries about our universe to date. Radio astronomy is said to have begun in 1932, when a man named Karl Jansky inadvertently figured out the source of a peculiar radio signal he was receiving. He was an engineer for Bell Laboratories, and he was grappling with an unusual static signal that was interfering with his shortwave radio transatlantic voice communications. After several months of trying to figure out the source of this interference, he noticed that it appeared to be moving across the sky. As this wasn't his area of expertise, he decided to explain what he was observing to a professional astronomer named Melvin Skellett. Karl Jansky wrote:

> *"I have taken more data which indicated definitely that the stuff, whatever it is, comes from something not only extra-terrestrial, but from outside the solar system. It comes from a direction that is fixed in space and the surprising thing is that ... it is in the direction towards which the solar system is moving in space. According to [Melvin] Skellett ... there are clouds of 'cosmic dust' in that direction ..."*

Jansky had unwittingly stumbled upon something occurring in the heart of the Milky Way galaxy. He later wrote one of the most important papers in the history of astronomy, called "Radio Waves from Outside the Solar System." It was published in 1933 and launched the science of radio astronomy.

The published abstract from that paper reads as follows:

> *"In a recent paper on the direction of arrival of high-frequency atmospherics, curves were given showing the horizontal component of the direction of arrival of an electromagnetic disturbance, which I termed hiss type atmospherics, plotted against time of day. These curves showed that the horizontal component of the direction of arrival changed nearly 360° in 24 hours and, at the time the paper was written, this component was approximately the same as the azimuth of the sun, leading to the assumption that the source of this disturbance was somehow associated with the sun."*

The reference to the sun is important, as we will see.

We, as humans, have advanced to the point where we are able to launch radio signals and satellites into outer space, holding encrypted mathematical patterns—messages intended to give clues to ETs about who we are, where we are, and what we are all about. Those messages are mostly in the form of prime number patterns and complicated binary codes that unravel to describe complex things about humans and what we think we know about our solar system.

We have put those messages into radio signals and on satellites in the hope that other intelligent life will find the information, decode it, understand it, and then try to communicate with us. Our sending of encrypted messages out into the universe is well documented. We are launching prime numbers and binary code out into space in specific patterns, including complex information such as the binary description of our double helix DNA. If ETs find it, we assume ETs will figure it out, because we assume that mathematics is the universal language and ETs are super intelligent. We assume that their deductive reasoning skills are as advanced as their technological skills. We essentially send them mathematical puzzles to figure out based on our assumption of their ability to do so. One example is the golden record message launched by NASA on the Voyager satellite in 1977. You will probably agree that the message would be very complicated to figure out if it was ever actually found by an ET.

The following images and description of the contents of the Voyager message is taken from NASA's Jet Propulsion Laboratory, California Institute of Technology website:

> *The contents of the record were selected for NASA by a committee chaired by Carl Sagan of Cornell University, et al. Dr. Sagan and his associates assembled 115 images and a variety of natural sounds, such as those made by surf, wind and thunder, birds, whales, and other animals. To this they added musical selections from different cultures and eras, and spoken greetings from Earth-people in fifty-five languages, and printed messages from President Carter and UN Secretary General Waldheim.*
>
> *"The spacecraft will be encountered and the record played only if there are advanced spacefaring civilizations in interstellar space."*
>
> —*Carl Sagan*

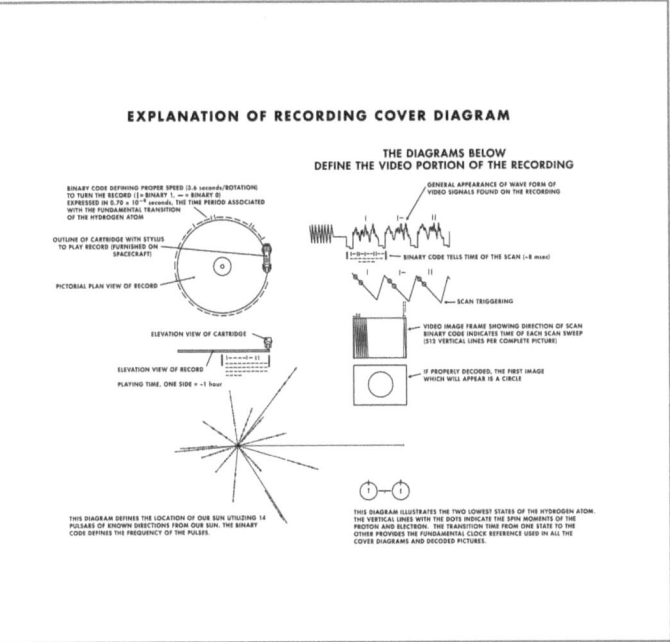

Each record is encased in a protective aluminum jacket, together with a cartridge and a needle. Instructions, in symbolic language, explain the origin of the spacecraft and indicate how the record is to be played. The 115 images are encoded in analog form.

The remainder of the record is in audio, designed to be played at 16-2/3 revolutions per minute. It contains the spoken greetings, beginning with Akkadian, which was spoken in Sumer about six thousand years ago, and ending with Wu, a modern Chinese dialect. Following the section on the sounds of Earth, there is an eclectic ninety-minute selection of music, including both Eastern and Western classics and a variety of ethnic music. Once the Voyager spacecraft leave the solar system (by 1990, both will be beyond the orbit of Pluto), they will find themselves in empty space. It will be forty thousand years before they make a close approach to any other planetary system. As Carl Sagan has noted, "The spacecraft will be encountered and the record played only if there are advanced spacefaring civilizations in interstellar space. But the launching of this bottle into the cosmic ocean says something very hopeful about life on this planet."

The definitive work about the Voyager record is "Murmurs of Earth" by Executive Director, Carl Sagan, Technical Director, Frank Drake, Creative Director, Ann Druyan, Producer, Timothy Ferris, Designer, Jon Lomberg, and Greetings Organizer, Linda Salzman. Basically, this book is the story behind the creation of the record, and includes a full list of everything on the record. "Murmurs of Earth," originally published in 1978, was reissued in 1992 by Warner News Media with a CD-ROM that replicates the Voyager record. Unfortunately, this book is now out of print, but it is worth the effort to try and find a used copy or browse through a library copy.

What do you think? Perhaps that is too complicated a message to figure out—if it is ever found at all. Voyager 2 got to Jupiter in July 1979 and Saturn in August 1981. So far as the general public knows, there has been no obvious, credible response to any of our messages. If our assessment of ETs is correct, then there could be several reasons for this lack of response. The three most obvious possibilities are:

1. The format of the messages is too complicated.
2. Our messages are not getting to where they need to go and not being received.
3. We are sending the wrong messages.

What if we flip the situation around and assume ETs sent us a message puzzle to figure out? If, as we assume, ETs left us a coded message based on the concept of three-way communication, they are expecting us to find their message and repeat it back to them, and then they will confirm with a response. Perhaps we are sending the wrong messages at the wrong times—messages that are too complicated—and we are sending them in the wrong direction.

In what form would ETs have left us a way to make contact, or a map to show us how they got here and where they go when they leave? What would it look like? It would probably not be a paper map where X marks the spot.

Their message is more likely to be a program or code to first discover, then decipher in order to make contact. It is not enough to dig up artifacts and study

It is reasonable to expect that ETs have left us clues to point us in the right direction

megalithic structures and speculate on what they all mean. Those things are all solid clues, but they yield no specific answers to the problem of how to make contact and establish communication.

The code provided by ETs has to be in a form that we can only figure out when we've advanced far enough to get into the space travel and/or space communication game. Then, ETs can be satisfied that because we've figured out the code, we must be similar enough to them to handle open communication. We will not run into a cave and hide, or worship them as gods (again) if they appear before us, ready to communicate. We are at the point now where most people believe ETs exist in some form. Limited and selective contact has been made, and they are coming and going all the time. But they haven't yet sent an open and direct communication for the entire world to see.

Maybe they are just being cautious. It is reasonable to expect that ETs have left us clues to point us in the right direction and that they require us to put the clues together—just as we are doing to them. To get to the map that opens the door, we need to recognize and find the keys. It is reasonable to expect that the map they've

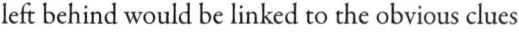

left behind would be linked to the obvious clues in the form of megalithic structures lining up with the constellation Orion or other celestial bodies. It is also reasonable to expect that such a map would use digital and/or mathematical media to create, store, and protect it until such time as we become advanced enough to discover and use those types of clues.

But what are the clues? We need to look for the obvious and the simple things first. It is reasonable to assume, as we already have, that the keys to interstellar communication with ETs is somehow linked to mathematics as the "universal language," and, more specifically, involves prime numbers and the mathematics of computer language, which is binary code. They would probably have put limiting parameters around their message to keep it simple and keep us on the right track. It is reasonable to assume that the most basic boundary would probably be the prime number 3 in conjunction with the first three prime numbers, 2, 3, and 5. On the surface, that would seem to be as simple as it gets for a starting place.

Why start with the prime number 3? It has been said that once is chance, twice is a coincidence, and a third time is a pattern. There are numerous and conspicuous examples of how the number 3 turns up as a symbol and number of importance over and over again in history, mathematics, religion, mythology, and virtually everywhere else. The conspicuous occurrences of the number 3 are a well-known subject that has been written about by more than one researcher. Some of the many examples include:

1. Water is the key to life (as we currently know it), and we spend a lot of time and money looking for evidence of water on other planets. Water is a three-atom compound: H_2O.

2. Some of the more significant Bible references are:
   a. The Holy Trinity (Father, Son, and Holy Ghost).
   b. The Three Wise Men who traveled to Bethlehem when Jesus was born, bringing three gifts of gold, frankincense, and myrrh. All of these gifts have important symbolic meanings.

    c. Jesus rose from the dead on the third day.
    d. There are three angels encountered in the Book of Revelation.
    e. Peter denies Jesus three times during Jesus's capture and crucifixion.

3. The Earth is the third planet from our sun.

4. The average human brain weighs approximately three pounds.

5. There are three primary parts of the human brain.

6. Ancient Babylonians believed in three primary gods: Anu, who represented Heaven; Baal, who represented Earth; and Ea, who represented the Abyss.

7. The Egyptian sun god, Ra, was their most powerful god, and it was believed he created everything. There were three aspects of Ra. He was Khepri when rising, Ra (Re) during midday, and Atum when setting.

8. In numerology, the number 3 is linked to the Ascended Masters. Numerologists believe that the Ascended Masters are enlightened spiritual "beings" who teach, inspire, protect, and guide the human race. The belief in numerology has been around from as far back as AD 325, and it is defined by Wikipedia as follows:

> *Numerology is any belief in the divine or mystical relationship between a number and one or more coinciding events. It is also the study of the numerical value of the letters in words, names, and ideas. It is often associated with the paranormal, alongside astrology and similar divinatory arts.*

    There is a lot more to it than that, but that is the premise.

9. Hinduism also has a holy trinity, called the Hindu triumvirate: Brahma the creator, Vishnu the preserver, and Shiva the destroyer.

10. The number 3 is found throughout Greek and Roman mythology, for example:

Three brothers ruled in Greek mythology:
a. Zeus—ruler of the sky
b. Poseidon—ruler of the sea
c. Hades—ruler of the underworld

11. The third eye (taken from Gaia.com, October 26, 2019):

> *The pineal gland is a pea-sized gland shaped like a pine cone, located in the vertebrate brain near the hypothalamus and pituitary gland. Also known as the third eye, it is a revered tool of seers and mystics and considered to be the organ of supreme universal connection. Its significance appears in every ancient culture throughout the world. For example, in Ayurvedic philosophy, the third eye is represented by the Ajna chakra and in Ancient Egypt, the symbol of the Eye of Horus mirrors the placement of the pineal gland in the profile of the human head. The third eye is connected to clarity, concentration, imagination and intuition.*
>
> *The pineal gland represents the third eye in biology, which produces melatonin. Melatonin controls circadian rhythms and reproductive hormones. This makes the pineal a master regulator of time, affecting not only our sleep patterns but also our sexual maturation. Melatonin also affects our stress and ability to adapt to a changing world. This third eye activates when exposed to light, and has a number of biological functions in controlling the biorhythms of the body. It works in harmony with the hypothalamus gland which directs the body's thirst, hunger, sexual desire and the biological clock that determines our aging process.*
>
> *Developing the third eye is the doorway to all things psychic—telepathy, clairvoyance, lucid dreaming and astral projection. The illusion of separation between self and spirit*

dissolves when the third eye connection is cultivated. Metaphysical ways of being are connected to the third eye, such as how to be awake within the dream, to walk between realities and surpass the limitations of humanity.

The conspicuous occurrences of the number 3 are a well-known and well-documented subject, but nobody knows how it all ties together—perhaps until now, as we will soon see.

The number 33 also has significance in religions and mythology from all over the world in ancient times through today. One coincidence that is not so obvious but that is directly applicable to the subject of this book is the presidency of Harry Truman. Truman was the 33rd president of the United States—the number 3 again. He was also (among other things):

1. A Freemason for over fifty years and was elected as Grand Master Mason of the Missouri Lodge.

2. A member of Shriners International, which is a Masonic society established in 1870. The society was originally named the "Ancient Arabic Order of the Nobles of the Mystic Shrine."

3. A member of the Sons of Confederate Veterans.

He was a true patriot in every sense of the word and had numerous accomplishments, all of which supported and honored the success of United States citizens and his country. He loved his country and was committed to its defense from all enemies, foreign and domestic—and, as it would seem, from ETs.

Truman was the only president who served in combat on foreign soil in WORLD WAR I. The other presidents who served in WORLD WAR I were Howard Taft (Connecticut Home Guard), Woodrow Wilson (President), Franklin D. Roosevelt (Assistant Secretary to the Navy), and Dwight D. Eisenhower (Stateside Service).

The National Archives (record Group 407) on Truman reads as follows:

Harry S. Truman enlisted for service in WORLD WAR I with the National Guard and received his commission as a first lieutenant, Battery F, 2nd Field Artillery Regiment, Missouri National Guard, on June 22, 1917. On September 5, 1917, Truman's regiment was called into federal service as the 129th Field Artillery Regiment. Truman was promoted to captain on April 23, 1918. He commanded Battery D of the 129th Field Artillery in France during the War. Truman was honorably discharged on May 6, 1919. He received his commission as a major in the Officers' Reserve Corps in 1920. Truman remained an officer in the Field Artillery Reserve until he retired with the rank of colonel on January 31, 1953.

Truman was a combat soldier who fought for his life in the field and directly saw, lived, smelled, tasted, and survived the horror of war, and the daily threat of a horrible death. He was also the president who authorized the first and only use of atomic bombs on civilians in WORLD WAR II as a weapon of mass destruction in August 1945. He did this in his seventh month as president, and the significance of the number 7 will be discussed later in this book. Truman knew how horrible the result would be for the victims. He also knew it would be the end of WORLD WAR II. How does a man make that decision without some sort of divine intervention and help? There is no way that any rational person can make that decision alone, or by relying only on opinions from presidential "advisors." Perhaps ETs were involved in influencing that decision?

One year later, in 1946, Truman created the Atomic Energy Commission. This did many things, but perhaps the most important thing it did was to transfer the control of the development of nuclear power from military to civilian hands.

Perhaps the dropping of nuclear bombs finally got the attention of ETs, and they started paying more attention to our activities here on Earth. In 1947, only two years after the atomic bomb drop, came the infamous Roswell, New Mexico "alleged" crash of a UFO that remains an unsolved mystery to the general public. To this day, the government will not admit that it was a UFO crash. It was also in 1947 that Truman established the Central Intelligence Agency (CIA). He then established the National Security Agency (NSA) as the organization within the US government that is charged with the responsibility of communications intelligence. The National Security Act of 1947 established the United States Department of the Air Force as its own branch of the United States Armed Services. It was President Truman who "allegedly" ordered the creation of the government agency called the "Majestic Twelve" (MJ-12) soon after the Roswell crash. It is claimed that  the MJ-12 group was comprised of important military personnel and leading scientists who were charged with the task of finding, recovering, and investigating ET spaceships. The government claims that this organization does not and never did exist.

In 1948, there was an official government project launched called "Project Sign". The mission of Sign was to collect and assess any and all information relating to UFO sightings that could be a threat to national security. After Sign, a similar government project called "Grudge" was created that quickly morphed into "Project Blue Book," which began in 1952. Blue Book had pretty much the same directive as its predecessors, investigating over 12,000 UFO reports. In all cases, their conclusion was that none of them were ET vessels and that the sightings could be explained away. Over 12,000 reports and all were resolved beyond any shadow of a doubt? That is difficult to believe. The 33rd president of the United States of America was the only president to see combat in World War I; the man who authorized the use of atomic bombs; the man who created government ET agencies just after the Roswell, New Mexico incident… And we are expected to believe the government when they deny they have evidence of ETs? That is very difficult to swallow. What is believable is that President Truman has been brought to our attention by his link to the number 3, and as we investigate his legacy, much of it is linked to ET investigations, secrecy, and security from this potential threat. Truman clearly knew something about ETs.

In 2010, there was a press conference held in Washington, DC by Robert Hastings (an author writing on the topic of UFOs) and several former US military airmen. They claimed that the US Air Force was concealing the truth about the national security risk posed by UFOs at various nuclear bases. They claimed to have proof. They said that since 1948, ETs in spaceships have not only been visiting Earth, but hovering over nuclear missile sites and deactivating these missiles. Hastings and his entourage of airmen (credible people) believe that Earth is being visited by beings from another world and that those ETs are actively interfering with the potential use of nuclear weapons on Earth.

* * *

There are clues other than the number 3 that could lead us to the solution to the code. Look at the two primary shapes in geometry: the equilateral triangle and the circle. These two fundamental shapes are conspicuous everywhere throughout history and all over the world as important and meaningful symbols. They are also the most prevalent shapes that UFO observers report.

The circle is the only defined geometric shape that is a single line that begins where it ends and has one radius and one diameter. The Alpha and the Omega, if you will. It is also the only shape in geometry that has a component that seemingly never ends. That component is Pi: its circumference divided by its diameter.

The equilateral triangle has some unique properties compared with the other forms of triangles in that:

1. All three sides and all three angles are equal.
2. It is the strongest structural engineering geometric shape.
3. The angle bisectors, the medians, and the perpendicular bisectors of the three sides coincide at a point. That point is the centroid of the triangle.
4. The incenter and the circumcenter coincide at the same point within the triangle. That point is the centroid of the triangle.

Based on some very basic boundaries selected because they have meaning, we can now begin to think about constructing a model using these boundaries.

1. The number 3
2. The first 3 prime numbers: 2–3–5
3. Binary code
4. The circle
5. The equilateral triangle

The Eye of Providence, or "All-Seeing Eye," is an ancient circle-and-triangle symbol used by numerous cultures throughout history. The circle and triangle mean something to humanity with respect to ET links and must somehow be involved in our search for an ET message. The image to the right is the all-seeing eye on the US one-dollar bill.

4.

The Model

ETs would have made the key to the code simple enough that we would have a reasonable shot at building a way to decode their message; they would want us to figure it out. It also has to be complex enough to require a certain level of intellectual and mathematical skill worthy of getting the attention of the ETs and being introduced to the community of the universe. As a third criterion, they may require that solving the code takes more than just mathematical muscle. A certain level of deductive reasoning skills may be needed, more than just the ability to solve an equation, so that they know the response we ultimately send is not coming from an artificial intelligence. We do the same thing when doing certain tasks online that require us to click the box that says, "I am not a robot" before we can take the next step.

As a starting point, our process uses the first three prime numbers to build a binary code block that looks like this.

2=	0	0	1	1	0	0	1	0
3=	0	0	1	1	0	0	1	1
5=	0	0	1	1	0	1	0	1

Step 1: With the understanding that the number 3 is involved as a tool to construct the model, we repeat the process 3 times vertically, rotating the position of the prime numbers in the binary block above 3 times such that each of the prime numbers occupies each vertical position at least one time. Note that the color code for each prime number remains consistent throughout this process. It helps to identify the pattern. The result is this:

2=	0	0	1	1	0	0	1	0
3=	0	0	1	1	0	0	1	1
5=	0	0	1	1	0	1	0	1
3=	0	0	1	1	0	0	1	1
5=	0	0	1	1	0	1	0	1
2=	0	0	1	1	0	0	1	0
5=	0	0	1	1	0	1	0	1
2=	0	0	1	1	0	0	1	0
3=	0	0	1	1	0	0	1	1

Each prime number appears at the top, the middle, and the bottom of the original 3-line block in sequence one time.

Step 2: Repeat the process in the horizontal direction and the model looks like this:

	2								3							5							
2	0	0	1	1	0	0	1	0	0	0	1	1	0	0	1	1	0	0	1	1	0	1	0	1
3	0	0	1	1	0	0	1	1	0	0	1	1	0	1	0	1	0	0	1	1	0	0	1	0
5	0	0	1	1	0	1	0	1	0	0	1	1	0	0	1	0	0	0	1	1	0	0	1	1
3	0	0	1	1	0	0	1	1	0	0	1	1	0	1	0	1	0	0	1	1	0	0	1	0
5	0	0	1	1	0	1	0	1	0	0	1	1	0	0	1	0	0	0	1	1	0	0	1	1
2	0	0	1	1	0	0	1	0	0	0	1	1	0	0	1	1	0	0	1	1	0	1	0	1
5	0	0	1	1	0	1	0	1	0	0	1	1	0	0	1	0	0	0	1	1	0	0	1	1
2	0	0	1	1	0	0	1	0	0	0	1	1	0	0	1	1	0	0	1	1	0	1	0	1
3	0	0	1	1	0	0	1	1	0	0	1	1	0	1	0	1	0	0	1	1	0	0	1	0

We now have 3 binary blocks of the first 3 prime numbers.

Step 3: Repeat it all 3 times and you end up with a model that looks like this:

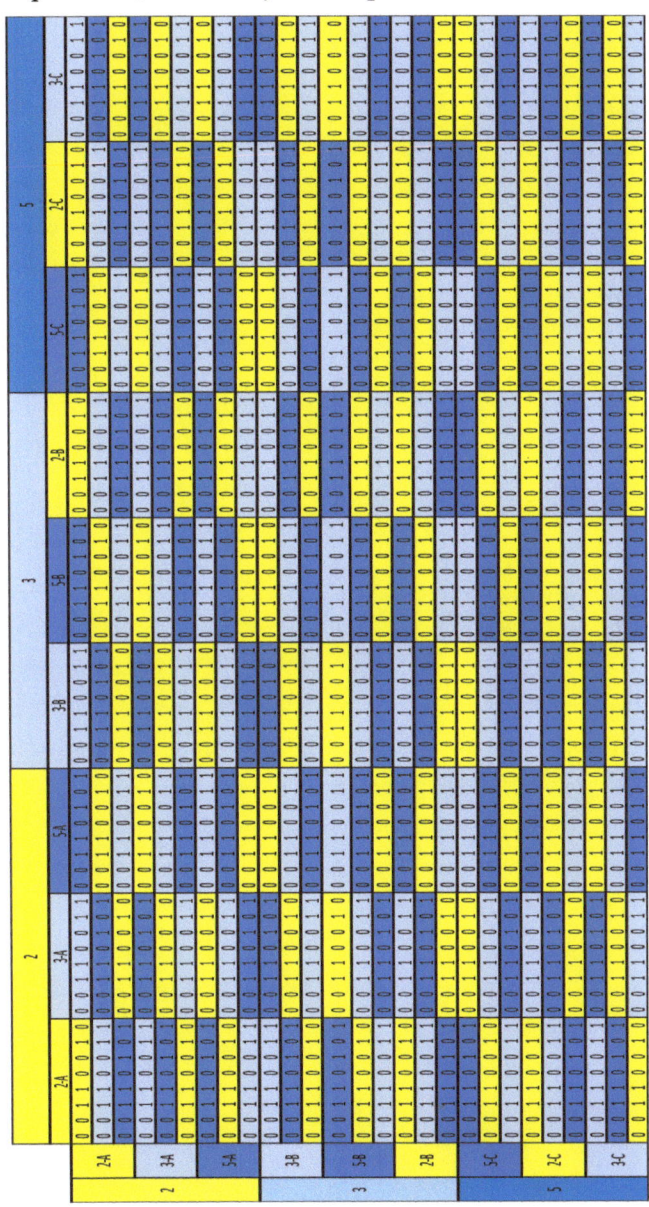

This result takes the first 3 prime numbers in a block and maneuvers them through a 3-step process to get a 3 × 3 block of the first 3 prime numbers rotated 3 times in a way that does not repeat. That's a lot of usage of the number 3, and this is the dream/vision that I discussed at the start.

This block of binary numbers was built by using these 3 keys to unlocking the code:

1. The number 3
2. The first 3 prime numbers: 2–3–5
3. Binary code

It's important to understand the concept of this process. It is a very similar thing to what we did when we sent messages to ETs. We developed a message of words converted into binary code to launch into the universe. ETs may also have deduced that this is the best way to open communications. They may have had the same con-

Perhaps we have to prove ourselves worthy of an ET response by figuring out how to retrieve the message

cept in mind and left us a binary coded message to decipher. The difference is in the mechanics of how to decode the message. We are sending our messages in the form of a puzzle to figure out before they get to the relevant information. Maybe ETs are doing the same thing to us—giving us a puzzle to figure out before we get to the message. Perhaps we have to prove ourselves worthy of an ET response by figuring out how to retrieve the message, then repeat the message or "instructions" back to them as Step 2 of the three-way communication model. Maybe that is the thing we have been missing all these years. This is a thought process and methodology that has never before been conceived or tested.

What are the next logical steps ETs might expect us to take if we got this far? Because right now, our giant block of ones and zeros is telling us nothing. It is reasonable that ETs would expect us to try to all the numbers into a kind of language that is not mathematical. That is exactly what we are doing when we send binary code messages out into the universe. Our codes translate into letters and words and diagrams and sounds, etc. Our messages are complicated to decipher. If ETs are similar to us in the way we think, as previously discussed, they may have been thinking in a similar way with respect to sending messages, except their message is much less complex—or at least different.

Take the Arecibo message, for example. The Arecibo radio telescope is located in Puerto Rico. It is a significant piece of equipment that was built in the early 1960s. In 2020, it fell apart and there are no plans to reconstruct this tool. Its useful life is over. Over the years, it has been used by astrophysicists to document many important observations and discoveries about our universe. It was the largest and most sensitive single-aperture radio telescope in the world until 2016, when China finished the construction of their FAST 500M radio telescope. An image of Arecibo's 1,000-foot diameter dish taken from their website is shown below:

There is much literature documenting the 1972 project when Arecibo sent a coded message into space in hopes of making contact with intelligent life out there. The message consisted of seven encoded parts as follows:

1. The numbers one through ten.
2. The atomic numbers of the elements for hydrogen, carbon, nitrogen, oxygen, and phosphorus, which make up our DNA.
3. The formulas for the sugars and bases in the nucleotides of our DNA.
4. The number of nucleotides in our DNA and a graphic of the double helix structure of our DNA.
5. A graphic figure of a human, the dimension (physical height) of an average man, and the human population of Earth.
6. A graphic of our solar system indicating from where the message was generated.
7. A graphic of the Arecibo radio telescope itself.

The message is shown below and can be found on numerous websites. It is widely published. It is very complex, and it is difficult to believe that if an ET ever found it, they would figure it out.

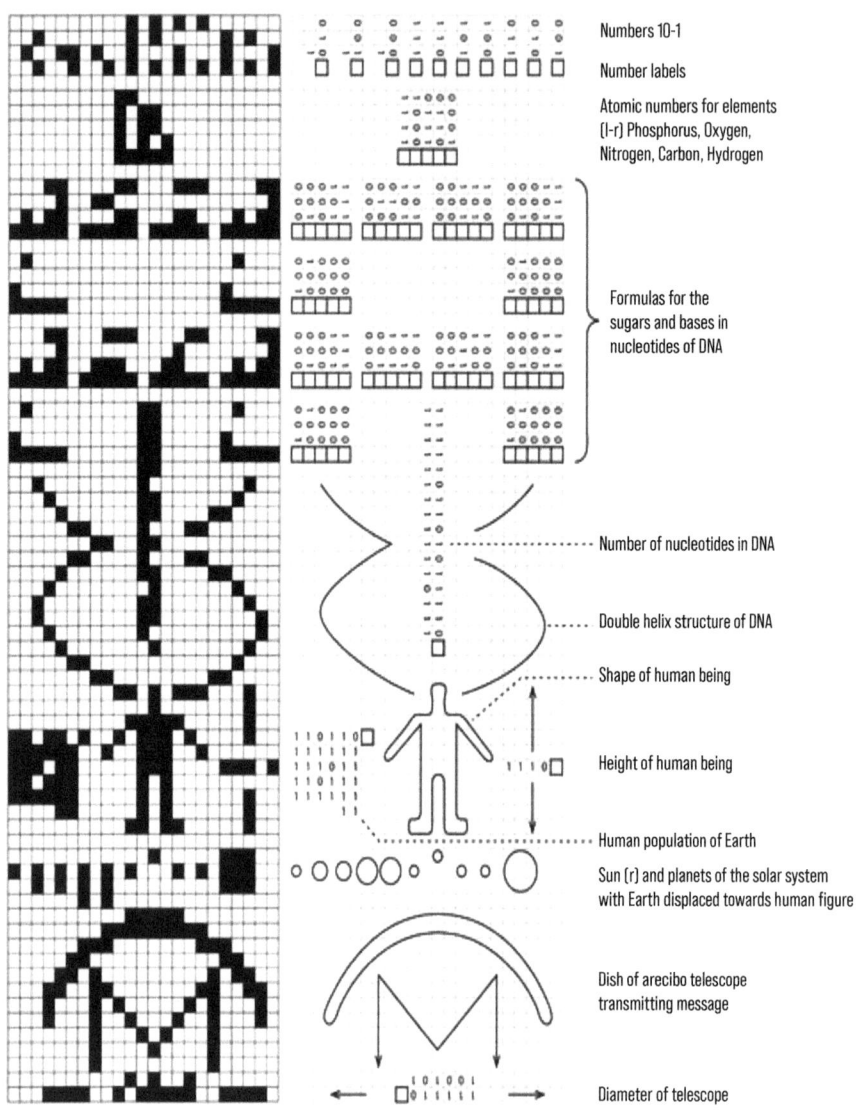

The binary code for all of this looks like this:

0000001010101000000000
0010100000101000000100
1000100010001001011010
1010101010101001001000 0000110000110000010000
0000000000000000000000 0001000000100000001000
0000000000011000000000 0010000000110000000100
0000000000110100000000 0100000000110000000100
0000000000110100000000 0100000000010000001000
0000000000101010000000 0010000000100000010000
0000000001111100000000 0001000000000000110000
0000000000000000000000 0000110000000011000000
110000111000110000110 00 0010001110101100000000
100000000000011001000 0 0010000000100000000000
110100011000110000110 10 0010000011111000000000
1111101111011111011 11 0010000101110100101101
0000000000000000000000 0000001001110010011111
0001000000000000000010 1011100011100000110111
0000000000000000000000 0000000001010000011101 1
0001000000000000000001 0010000010100000111111
1111100000000000011111 0010000010100000110000
0000000000000000000000 0010000110110000000000
110000110001110011000 0000000000000000000000
100000010000000010000 0011100001000000000000
110100011000111001101 0 0011101010001010101010 1
1111101111011111011 11 0011100000000101010100
0000000000000000000000 0000000000000101000000
0001000000110000000010 0000000011110000000000
0000000000110000000000 0000001111111100000000
0001000001100000000001 0000111000000111000000
1111100001100000011111 0001100000000001100000
0000000000110000000000 0011010000000001011000
0010000000010000000100 0110011000000110011000
0001000000110000001000 0100010100001010001000
0000110000110000010000 0100010010010010001000
0000011000100001100000 0000010001010001000000
0000000000110011000000 0000010000100010000000
0000001100010000110000 0000010000000010000000
                         0000000100101000000000
                         0111100111101001111000

We ourselves would probably have a hard time figuring out this message, and we created this beast.

The message contained 1,679 binary digits configured in 73 columns and 23 rows and took less than 3 minutes to transmit. The message was directed at the M13 cluster, largely because it is a big and relatively close (approximately 25,000 light-years away) collection of stars that was available at the time of transmission. M13 lies within the Hercules constellation.

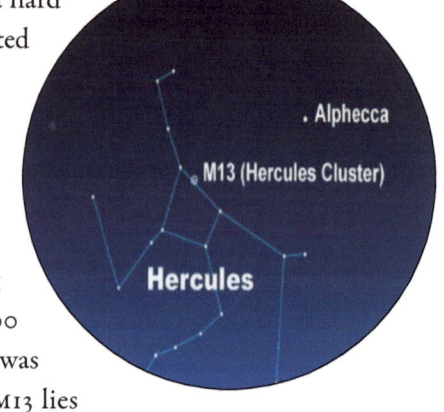

That seems like a very random way to decide where to send a fairly important and complicated message that, if received and returned, would change the world. The cluster was discovered in 1794 by a man named Edmund Halley. A Hubble image of M13 is shown below.

The M13 cluster is approximately 145 light-years in diameter. It is visible in the northern hemisphere in the spring and summer. The Arecibo message will take approximately 21,000 years to reach M13, and it would take the same amount of time, if returned by the same method, for a response to reach Earth. In addition, since the launch of the Arecibo message, astronomers are now of the belief that these sorts of dense clusters do not often include planets, habitable or not. The Arecibo message can't be considered a genuine effort to communicate. It would seem that the Arecibo message was more intended to demonstrate human

technological advancement than as a genuine and targeted attempt to start a dialog with intelligent life in the universe.

The difference between what we are doing here and the Arecibo message is that we are trying to figure out an existing message and repeat it back. We are not trying to construct and send an original message to ETs. The Arecibo message sent out some fairly complicated and difficult-to-decipher concepts in a more-or-less random direction of convenience. We are looking for an existing message that is simple to understand, with the aim of returning it in a justified/guided/mapped direction.

Now, getting back to the binary block analysis, if we go column by column, picking off binary sequences that translate into letters, the result is this:

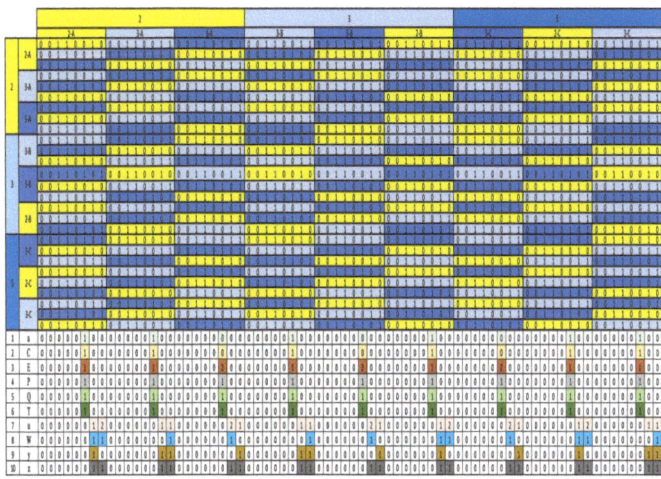

These are the ten letters that pop out of the vertical analysis:

a, C, E, P, Q, T, u, W, y, z

This method of analysis is different from a typical search for words and letters in a binary block. It does not read like a book. The analysis was done by looking for all the letters of the alphabet in uppercase and lowercase, as the binary values are different for each case. Notice that within each binary number column block (2-A, 3-A, 5-A, etc.), the first 5 columns yield no letters. Letters only appear in the last 3 columns of each prime number block. This is our first occurrence of the number 3 yielded by our process, which uses 3 as a boundary condition. Also notice that in the columns that do produce letters, multiple letters are produced. This is because each column was tested for every letter of the alphabet. There is an overlap in letter "hit" sequences. For example, if we look at the letters "A" and "C," their respective binary numbers are:

A = 01100001
C = 01000011

The common/overlap digits are highlighted in yellow. When we take this to the binary block, it looks like this:

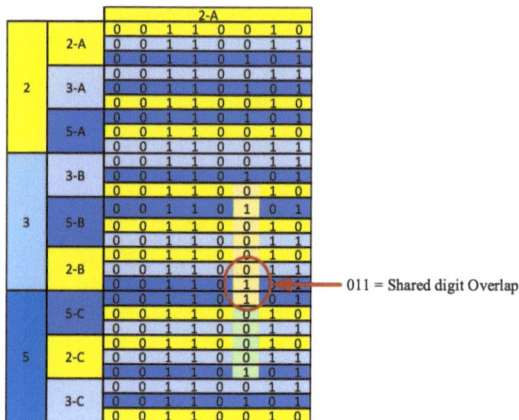

To get our results, each column and row was tested for every letter of the alphabet. We did not start at the top, find a letter, and then keep going. We examined the entire block for one letter, then "cleaned the slate" and examined the entire block for the next letter. Then we dealt with the results in a way that will be explained. This is a very different approach than the traditional way of reading a block of code. This is deliberate, as we are trying to "think differently," as discussed earlier. Our block is a 3-tiered binary block with 72 columns and 27 rows. This is a 1,944-digit block. The Arecibo block was a 1,679-digit message (265 digits smaller). Based on that, one would expect anything that comes out of our block to be at least as complicated as the Arecibo block; that may or may not be the case. You can be the judge of complexity in the end. The important difference is the methodology used to interpret our block.

We can see that in alphabetical order (the far-left column), C is in the number 2 spot, E is in the number 3 spot, and Q is in the number 5 spot.

Recall that two of our keys are the first 3 prime numbers—2–3–5—and the number 3 alone, as they appear throughout history as numbers of significance. The letter match above is CEQ. Phonetically, that is pronounced "seek you." Eliminate the "E" that corresponds to the number 3 from our analysis and you are left with the letters CQ, which is also phonetically pronounced "seek you." The call-out letters "CQ" on amateur/HAM radio are used when the radio operator is asking anybody out there who is listening to respond.

Could this be the next clue our process has revealed? Are the ETs trying to indicate in this message that they "seek us" and that the contact will be by radio signal—like the way we discover pulsars? Is this the clue that tells us we are on the right trail? Is there a link between radio signals and the 2–3–5 sequence right out of the gate?

According to the *Astrophysical Journal* (October 10, 2018), Radio Pulsar PSR J0250+5854 was recently discovered, and it is the slowest-spinning pulsar known to date. It spins at the rate of 23.5 (2–3–5) seconds per spin. Recall that information about pulsars was part of the message we put on Voyager's golden record back in the 1970s. Did ETs also recognize the significance of pulsars as some form of "common ground" for trying to communicate with humans? Pulsars are neutron stars that are the remains of dead giant stars. They rotate at very fast speeds and are highly magnetized. They also emit radio pulses, and that is how we find them. The first one was officially discovered in 1967. There are now over 2,000 known pulsars that are between 7 and 20 times the mass of our sun. The point here is that our 2–3–5 input model has returned a 2–3–5 clue/link/response. It has linked us to the slowest known pulsar, and pulsars are found by the radio signals they emit. We also launched information about pulsars out into space on a satellite, so we have already found some common ground between our message and the ET code.

We have been sending and receiving radio signals in an effort to communicate with ETs for decades. We are now, in fact, receiving signals in patterns, but have not yet figured out what those patterns mean—if they mean anything at all. We don't know where the signals are coming from, specifically, or if they are being generated by intelligent life. The universe is actually a noisy place with respect to radio signals. It is incredibly vast, so we need to direct our search in a particular direction in a more focused manner. But what is the focus? What is the target, and where is the target?

Further analyzing our 3 x 3 binary matrix horizontally, row by row, to pick out binary sequences that translate into letters, you will extract the letters d, i, F, f, and S.

All of the letters in alphabetical order from vertical and horizontal analyses of the 3 × 3 binary block yield fifteen letters as follows (note that 15 is the product of the second and third prime numbers used in our analysis (3 × 5 = 15)):

a, C, d, E, F, f, i, P, Q, S, T, u, W, y, z

Now, of course, this result raises the question: how could ETs in ancient times—or at any time—put together some coded message that results in letters from the English alphabet that mean anything to us today, or at any time in our history? The Bible and other ancient texts are full of references to God or gods speaking in plain language to all the peoples on Earth in all languages. One of many examples comes from the Bible: Acts (of the Apostles) 2:2–11:

> ² *And suddenly there came a sound from heaven as of a rushing mighty wind, and it filled all the house where they were sitting.*
> ³ *And there appeared unto them cloven tongues like as of fire, and it sat upon each of them.*
> ⁴ *And they were all filled with the Holy Ghost, and began to speak with other tongues, as the Spirit gave them utterance.*
> ⁵ *And there were dwelling at Jerusalem Jews, devout men, out of every nation under heaven.*
> ⁶ *Now when this was noised abroad, the multitude came together, and were confounded, because that every man heard them speak in his own language.*
> ⁷ *And they were all amazed and marveled, saying one to another, Behold, are not all these which speak Galileans?*
> ⁸ *And how hear we every man in our own tongue, wherein we were born?*
> ⁹ *Parthians, and Medes, and Elamites, and the dwellers in Mesopotamia, and in Judaea, and Cappadocia, in Pontus, and Asia,*
> ¹⁰ *Phrygia, and Pamphylia, in Egypt, and in the parts of Libya about Cyrene, and strangers of Rome, Jews and proselytes,*
> ¹¹ *Cretes and Arabians, we do hear them speak in our tongues the wonderful works of God.*

It would seem that there was no language barrier between humans and "God" or ETs in ancient biblical times, and there never has been. There is evidence to suggest that the ETs that came to Earth could communicate with the peoples of Earth in their own Earth language before the creation of an alphabet in any language. Furthermore, there is evidence to suggest that ETs have intervened in the physical and intellectual development of humans over the entire course of human history. Perhaps that intervention included some intervention in language development.

5.

Model Results

Running all of those letters through a word scrambler program to identify only the 3-letter word combinations (because 3 is our main driver), 194 combinations are returned from the 15 letters. Note that different word unscramble programs can produce slight variations on the list. For this analysis, 3 different word unscramble programs produced the results shown in this book, either exactly, or within two or three placements. All of the results were run through this process to get to the results that showed the most promise with respect to what is being discussed in this book. These results seem to have the most relevance.

194 is a lot of combinations to think about in an attempt to decipher some sort of simple code. Staying within the boundary conditions and the premise that the ETs' message would need to be simple and controlled by the number 3, we can drill down further into the list, again by using the first 3 prime numbers and the number 3.

If we take the alphabetical list of 3-letter words and acronyms and run their number position in the list through 3 divisions of three, it looks like this:

1	ACE	26	CUD	51	EFS	76	IFS
2	ACT	27	CUE	52	EFT	77	IST
3	ADS	28	CUI	53	ESQ	78	ITS
4	ADZ	29	CUP	54	ETA	79	PAC
5	AES	30	CUT	55	FAD	80	PAD
6	AFF	31	DIA	56	FAP	81	PAS
7	AFT	32	DAP	57	FAS	82	PAT
8	AID	33	DAW	58	FAT	83	PAW
9	AIS	34	DAY	59	FAY	84	PAY
10	AIT	35	DEF	60	FED	85	PEA
11	APE	36	DEW	61	FET	86	PEC
12	APT	37	DEY	62	FEU	87	PES
13	ASP	38	DIE	63	FEW	88	PET
14	ATE	39	DIF	64	FEY	89	PEW
15	ATS	40	DIP	65	FEZ	90	PIA
16	AWE	41	DIS	66	FID	91	PIC
17	AYE	42	DIT	67	FIE	92	PIE
18	AYS	43	DUE	68	FIT	93	PIS
19	CAD	44	DUI	69	FIZ	94	PIT
20	CAP	45	DUP	70	FUD	95	PIU
21	CAT	46	DYE	71	ICE	96	PSI
22	CAW	47	EAT	72	ICY	97	PUD
23	CAY	48	EAU	73	IDE	98	PUS
24	CEP	49	ECU	74	IDS	99	PUT
25	CIS	50	EFF	75	IFF	100	PUY

3	2	1	Start	3	2	1	Start	3	2	1	Start
0.04	0.11	0.33	1	ACE	0.96	2.89	8.67	26	CUD	1.89	5.67	17.00	51	EFS
0.07	0.22	0.67	2	ACT	1.00	3.00	9.00	27	CUE	1.93	5.78	17.33	52	EFT
0.11	0.33	1.00	3	ADS	1.04	3.11	9.33	28	CUI	1.96	5.89	17.67	53	ESQ
0.15	0.44	1.33	4	ADZ	1.07	3.22	9.67	29	CUP	2.00	6.00	18.00	54	ETA
0.19	0.56	1.67	5	AES	1.11	3.33	10.00	30	CUT	2.04	6.11	18.33	55	FAD
0.22	0.67	2.00	6	AFF	1.15	3.44	10.33	31	DIA	2.07	6.22	18.67	56	FAP
0.26	0.78	2.33	7	AFT	1.19	3.56	10.67	32	DAP	2.11	6.33	19.00	57	FAS
0.30	0.89	2.67	8	AID	1.22	3.67	11.00	33	DAW	2.15	6.44	19.33	58	FAT
0.33	1.00	3.00	9	AIS	1.26	3.78	11.33	34	DAY	2.19	6.56	19.67	59	FAY
0.37	1.11	3.33	10	AIT	1.30	3.89	11.67	35	DEF	2.22	6.67	20.00	60	FED
0.41	1.22	3.67	11	APE	1.33	4.00	12.00	36	DEW	2.26	6.78	20.33	61	FET
0.44	1.33	4.00	12	APT	1.37	4.11	12.33	37	DEY	2.30	6.89	20.67	62	FEU
0.48	1.44	4.33	13	ASP	1.41	4.22	12.67	38	DIE	2.33	7.00	21.00	63	FEW
0.52	1.56	4.67	14	ATE	1.44	4.33	13.00	39	DIF	2.37	7.11	21.33	64	FEY
0.56	1.67	5.00	15	ATS	1.48	4.44	13.33	40	DIP	2.41	7.22	21.67	65	FEZ
0.59	1.78	5.33	16	AWE	1.52	4.56	13.67	41	DIS	2.44	7.33	22.00	66	FID
0.63	1.89	5.67	17	AYE	1.56	4.67	14.00	42	DIT	2.48	7.44	22.33	67	FIE
0.67	2.00	6.00	18	AYS	1.59	4.78	14.33	43	DUE	2.52	7.56	22.67	68	FIT
0.70	2.11	6.33	19	CAD	1.63	4.89	14.67	44	DUI	2.56	7.67	23.00	69	FIZ
0.74	2.22	6.67	20	CAP	1.67	5.00	15.00	45	DUP	2.59	7.78	23.33	70	FUD
0.78	2.33	7.00	21	CAT	1.70	5.11	15.33	46	DYE	2.63	7.89	23.67	71	ICE
0.81	2.44	7.33	22	CAW	1.74	5.22	15.67	47	EAT	2.67	8.00	24.00	72	ICY
0.85	2.56	7.67	23	CAY	1.78	5.33	16.00	48	EAU	2.70	8.11	24.33	73	IDE
0.89	2.67	8.00	24	CEP	1.81	5.44	16.33	49	ECU	2.74	8.22	24.67	74	IDS
0.93	2.78	8.33	25	CIS	1.85	5.56	16.67	50	EFF	2.78	8.33	25.00	75	IFF

3	2	1	Start	3	2	1	Start	3	2	1	Start
2.81	8.44	25.33	76	IFS	3.74	11.22	33.67	101	PYA	4.67	14.00	42.00	126	SIP
2.85	8.56	25.67	77	IST	3.78	11.33	34.00	102	PYE	4.70	14.11	42.33	127	SIT
2.89	8.67	26.00	78	ITS	3.81	11.44	34.33	103	QAT	4.74	14.22	42.67	128	SPA
2.93	8.78	26.33	79	PAC	3.85	11.56	34.67	104	QIS	4.78	14.33	43.00	129	SPY
2.96	8.89	26.67	80	PAD	3.89	11.67	35.00	105	QUA	4.81	14.44	43.33	130	STY
3.00	9.00	27.00	81	PAS	3.93	11.78	35.33	106	QUE	4.85	14.56	43.67	131	SUE
3.04	9.11	27.33	82	PAT	3.96	11.89	35.67	107	QUI	4.89	14.67	44.00	132	SUI
3.07	9.22	27.67	83	PAW	4.00	12.00	36.00	108	SAC	4.93	14.78	44.33	133	SUP
3.11	9.33	28.00	84	PAY	4.04	12.11	36.33	109	SAD	4.96	14.89	44.67	134	SUQ
3.15	9.44	28.33	85	PEA	4.07	12.22	36.67	110	SAE	5.00	15.00	45.00	135	SWY
3.19	9.56	28.67	86	PEC	4.11	12.33	37.00	111	SAI	5.04	15.11	45.33	136	TAD
3.22	9.67	29.00	87	PES	4.15	12.44	37.33	112	SAP	5.07	15.22	45.67	137	TAE
3.26	9.78	29.33	88	PET	4.19	12.56	37.67	113	SAT	5.11	15.33	46.00	138	TAI
3.30	9.89	29.67	89	PEW	4.22	12.67	38.00	114	SAU	5.15	15.44	46.33	139	TAP
3.33	10.00	30.00	90	PIA	4.26	12.78	38.33	115	SAW	5.19	15.56	46.67	140	TAS
3.37	10.11	30.33	91	PIC	4.30	12.89	38.67	116	SAY	5.22	15.67	47.00	141	TAU
3.41	10.22	30.67	92	PIE	4.33	13.00	39.00	117	SAZ	5.26	15.78	47.33	142	TAW
3.44	10.33	31.00	93	PIS	4.37	13.11	39.33	118	SEA	5.30	15.89	47.67	143	TEA
3.48	10.44	31.33	94	PIT	4.41	13.22	39.67	119	SEC	5.33	16.00	48.00	144	TED
3.52	10.56	31.67	95	PIU	4.44	13.33	40.00	120	SEI	5.37	16.11	48.33	145	TEW
3.56	10.67	32.00	96	PSI	4.48	13.44	40.33	121	SEQ	5.41	16.22	48.67	146	TFW
3.59	10.78	32.33	97	PUD	4.52	13.56	40.67	122	SET	5.44	16.33	49.00	147	TIC
3.63	10.89	32.67	98	PUS	4.56	13.67	41.00	123	SEW	5.48	16.44	49.33	148	TIE
3.67	11.00	33.00	99	PUT	4.59	13.78	41.33	124	SEZ	5.52	16.56	49.67	149	TIP
3.70	11.11	33.33	100	PUY	4.63	13.89	41.67	125	SIC	5.56	16.67	50.00	150	TIS

3	2	1	Start		3	2	1	Start	
5.59	16.78	50.33	151	TIZ	6.52	19.56	58.67	176	YAP
5.63	16.89	50.67	152	TUI	6.56	19.67	59.00	177	YAS
5.67	17.00	51.00	153	TUP	6.59	19.78	59.33	178	YAW
5.70	17.11	51.33	154	TWA	6.63	19.89	59.67	179	YEA
5.74	17.22	51.67	155	TYE	6.67	20.00	60.00	180	YEP
5.78	17.33	52.00	156	UPS	6.70	20.11	60.33	181	YES
5.81	17.44	52.33	157	USE	6.74	20.22	60.67	182	YET
5.85	17.56	52.67	158	UTA	6.78	20.33	61.00	183	YEW
5.89	17.67	53.00	159	UTE	6.81	20.44	61.33	184	YEZ
5.93	17.78	53.33	160	UTS	6.85	20.56	61.67	185	YIP
5.96	17.89	53.67	161	WAD	6.89	20.67	62.00	186	YIZ
6.00	18.00	54.00	162	WAE	6.93	20.78	62.33	187	YUP
6.04	18.11	54.33	163	WAP	6.96	20.89	62.67	188	YUS
6.07	18.22	54.67	164	WAS	7.00	21.00	63.00	189	ZAC
6.11	18.33	55.00	165	WAT	7.04	21.11	63.33	190	ZAP
6.15	18.44	55.33	166	WAY	7.07	21.22	63.67	191	ZAS
6.19	18.56	55.67	167	WAZ	7.11	21.33	64.00	192	ZES
6.22	18.67	56.00	168	WEI	7.15	21.44	64.33	193	ZIP
6.26	18.78	56.33	169	WET	7.19	21.56	64.67	194	ZIT
6.30	18.89	56.67	170	WEY					
6.33	19.00	57.00	171	WIS					
6.37	19.11	57.33	172	WIT					
6.41	19.22	57.67	173	WIZ					
6.44	19.33	58.00	174	WUZ					
6.48	19.44	58.33	175	WYE					

The result after the third round of division by 3 (column 3 on the left) is that there are 4 items (highlighted in green) that are divisible by 3 down to the first 3 prime numbers, 2–3–5, plus the 4th prime number, which is 7. Why does the number 7 (unexpectedly) drop out of the process and into our analysis?

The number 7 has significance in that 7 is the next prime number after the 2–3–5 binary block analysis that got us to this point (2–3–5–7). We put the 2–3–5 number sequence into the machine and turned the crank, and we got back words that correspond to 2–3–5 plus 7.

The clues being revealed are screaming out loud that we are definitely on the right trail.

The 7th letter of the Greek alphabet is Eta. It originally meant the number 7 too. Eta is also the first 3-letter group turned out by the model that corresponds above to the first prime number (2). The appearance of 7 seems to "close the loop." The Bible tells us that on the 7th day, God rested from creating the universe and all the life which it contains. From the Book of Genesis, chapter 2:

> [1] Thus the heavens and the earth were finished, and all the host of them.
> [2] And on the seventh day God ended his work which he had made; and he rested on the seventh day from all his work which he had made.
> [3] And God blessed the seventh day, and sanctified it: because that in it he had rested from all his work which God created and made.
> [4] These are the generations of the heavens and of the earth when they were created, in the day that the LORD God made the earth and the heavens,

In our code, the number 7 correlates to the 3-letter word Zac. Zac is an ancient Hebrew boy's name that means "The Lord has remembered." Is the 7 that shows up in our process at this point a clue that the code has ended on this last 3-letter item? The fact that the number 7 has dropped out of our process is significant. Recall from our discussion of President Truman that he was in the 7th month of his presidency when he dropped the atomic bombs on Hiroshima and Nagasaki. The end of WORLD WAR II and the end of the coded message in our analysis that could change the world are both linked to the number 7.

If we did a 4th round of division by 3 in our binary block, only the word Wae would appear in the prime number 2 position, and no other 3-letter word would be produced in any prime number positions. Wae is a Scottish term of sorrow ("woe")—"Woe is me." It's also (phonetically) what a rider says to their horse when they want it to stop. Clearly the process stops when the prime number 7 drops out after the 3rd round of division by 3.

The clues being revealed are screaming out loud that we are definitely on the right trail. Something is going on here. With the completion of each step in the process, we are given more clues that suggest we should keep going. The word search "filtering" results are:

2 = **ETA**

3 = **PAS**

5 = **SWY**

7 = **ZAC**

6.
Analysis of Result Words

NOTICE THAT THE SUM OF the prime numbers corresponding to the letter groups is (2 + 3 + 5 + 7 = 17). Not only is 17 a prime number, but it is also the 7th prime number. This is also a significant numerical relationship linked to the Bible and an important clue for this study. The number 17 is a significant number in Biblical terms because 17 symbolizes "overcoming the enemy" and "complete victory." God cleansed the earth of the failed and "unacceptable" humans by bringing the great flood, through rain, on the 17th day of the 2nd Hebrew month (Iyyar). In addition, the Bible says that Noah and his Ark safely rested on the mountains of Ararat on the 7th day of the 7th Hebrew month (Tishri). Notice the 17–7 links to our process output.

The 17th day of the 7th Hebrew month falls in the middle of the Christian Feast of the Tabernacles. It is the holiest month of the Jewish calendar.

The Feast of the Tabernacles is referenced throughout the Bible. It lasts for 7 days. One reference from the Book of Leviticus is as follows:

> *On the fifteenth day of the seventh month, when you have gathered in the produce of the land, you shall celebrate the feast of the LORD seven days. On the first day shall be a solemn rest, and on the eighth day shall be a solemn rest. And you shall take on the first day the fruit of splendid trees, branches of palm trees and boughs of leafy trees and willows of the brook, and you shall rejoice before the LORD your God seven days. You shall celebrate it as a feast to the LORD for seven days in the year. It is a statute forever throughout your generations; you shall celebrate it in the seventh month. You shall dwell in booths [sukkot] for seven days. All native Israelites shall dwell in booths [sukkot], that your generations may know*

that I made the people of Israel dwell in booths [sukkot] when I brought them out of the land of Egypt: I am the LORD your God (Leviticus 23:39–43 (English Standard Version)).

The parts that appear relevant to our process are:

- Sukkot is a happy 7-day holiday.
- It is connected to the harvest.
- It is an everlasting commandment.
- Jewish people in Israel are commanded to dwell in a sukkah for 7 days.
- It is a commemoration of the Hebrew exodus from slavery in Egypt.

This connection to our process makes a person wonder who or what, exactly, is leading us down this trail and why.

What started out as a mathematical matrix of the first 3 prime numbers that was analyzed using the number 3 as the main control has led us to 4 letter groups of 3 letters each. The first letter group corresponding to the first prime number (2) is Eta.

If you do some research into the various translations and definitions of Eta, a lot of interesting and relevant correlations can be made. A few of the more significant meanings are as follows:

1. **Greek alphabet Eta**

As noted previously, Eta is the seventh letter of the Greek alphabet. It was also originally the Greek number 7. The 7 brightest stars in the Orion constellation form an hourglass shape.

An hourglass is an ancient tool from the Middle Ages used to measure the passage of a certain amount of time. Orion—and, more specifically, Orion's belt—is known to have significance to many ancient cultures and the construction and positioning of their megalithic structures. We have already discussed the importance of the number 7 in our process so far. Much of what we know about geometry and mathematics comes from the ancient Greek mathematicians. The word mathematics itself is derived from the Greek *Mathema*, meaning the subject of instruction. We are using these principles (instructions) to guide us in a process to find our way to the ETs.

2. **International Morse Code Eta**

Much has been written about Samuel Morse, the inventor of International Morse Code, and his legacy will not be rewritten here, but he was a fascinating man. The simplistic overview as it applies to our subject is that Morse was the first person to standardize coded telecommunications sent down a wire that could be decoded at the other end. He knew he wanted to invent a code that could translate electrical signals into letters, and then words. That code had to be simple to understand and use. To simplify his code, he set out to determine the most frequently used letters of the alphabet. He did not go to the dictionary to count letters, but rather he counted the most frequently used letters in printers' type for common publications and news

> Morse was the first person to standardize coded telecommunications sent down a wire that could be decoded at the other end

bulletins, etc. This made sense because most of the words we use are simple and are repeated frequently in conversation and in printed communications.

Morse's search results were as follows, in order of frequency of use:

#1 = **E**
#2 = **T**
#3 = **A** (and also, I, N, O, and S)

Like mathematics, Morse code is a code of communication that is internationally recognized by all countries that have electronic communication capabilities. The code is as follows (taken from Wikipedia).

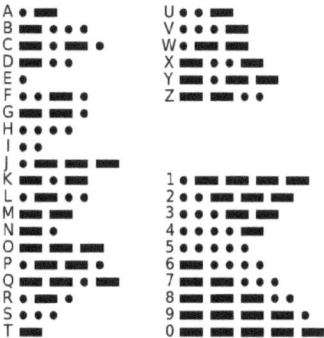

Let's take a closer look at this and note some relevant observations:

1. There are only 2 letters that have single symbols—the E (1 dot) and the T (1 dash).

2. If you put the letters E and T together, you get A (dot dash).

3. If you put the letters E, T, and A together, you get dot dash dot dash. This is the simplest alternating pattern possible. Recall that Eta is the first word that dropped out of our binary block analysis.

4. In keeping with our focus on the number 3, notice that there are only 2 letters that use a symbol 3 times alone with no other symbol—the S (dot dot dot) and the O (dash dash dash). Put these letters together in the simplest alternating pattern and we get SOS, the international distress signal. It means either "Save Our Ship" or "Save Our Souls."

Another relevant point is to do with the mechanics of how a standard message is transmitted in Morse code. A dot is considered 1 "unit" and a dash is considered 3 "units." When transmitting a message:

   a. The space between letters is 3 units.
   b. The space between words is 7 units.

The telegraph, the first form of telecommunication in the history of humankind, was operated using the International Morse code language and is linked to what we are doing here by means of the word Eta, the number 3, and the number 7. It is important to recognize that we were guided to Eta via a binary block of 3s and not by any counting of letter usage frequency in written or spoken language. The discovery of our Eta and Morse's Eta are not linked in that way. Our Eta and Morse's Eta were independently derived via different methodologies, but they have the same general intent—to simplify communications over long distances. That is called "independent verification."

3. Eta-Earth

There is a concept out there in the scientific/astrophysics world called Eta-Earth. The Encyclopedia of Astrobiology DOI 10.1007/978-3-642-27833-4_5306-4 # Springer-Verlag Berlin Heidelberg 2014; defines Eta-Earth as:

> *The term eta-Earth is defined as the mean number per star of rocky planets with between 1 and 1.5–2 Earth-radii that reside in the optimistic habitable zone (HZ) of their host star. Eta-Earth enters one formulation of the Drake equation, which endeavors to estimate the occurrence of intelligent life in the Galaxy; at the present time, it is usually calculated separately for each stellar spectral type. Thus, eta-Earth represents the occurrence rate of rocky planets in the optimistic HZ of different stars. The References present some values for eta Earth based on different statistical analyses of the data from the Kepler space telescope*

There is a lot more to it than is discussed here. The point is that Eta is a clue that relates to trying to figure out the potential locations of ET existences. Someone along the way called these potentially habitable planets "Eta-Earths." Perhaps an ET influence was involved in keeping the flow of clues coming. Note that NASA estimates that there are on the order of one billion "earths" in our galaxy alone, so the odds are that there's at least one other habitable planet out there capable of supporting intelligent life.

4. The Eta Function

In mathematics and astrophysics, there is a concept called the Eta function. It is linked to the study of string theory and the existence of wormholes.

A wormhole (also called an Einstein–Rosen bridge) is a theoretical "structure" in the universe that connects different points in spacetime. It is based on a special solution of Einstein field equations. A wormhole can be described as a tunnel with two ends at separate locations in space and/or time. The theory relies on Einstein's general theory of relativity, and the existence of wormholes has not been proven. In theory, wormholes can connect extremely long distances (as much as a billion light-years or more), short distances (such as a few feet), different galaxies, or different points in time.

Two of the questions to which we seek answers are: "How did ETs get here?" and "Where did/do they go?" You can't have that conversation without a discussion about wormholes. The Eta function is a tool used by some very smart people in search of an explanation, proof, and/or understanding of wormholes. It is now brought before us in this quest for answers via the word Eta that fell out of our binary block analysis; that link is significant.

5. Eta Carinae

Eta Carinae is a much-studied and gigantic stellar system that has two stars in relatively close proximity. Its mass is estimated to be 100–150 times the mass of our sun, and its luminosity is estimated to be about four million times that of our sun.

Eta Carinae lies approximately 7,500 light-years from Earth, within the constellation of Carina. Carina is Latin for the "keel" of a ship. Carina used to be part of a much larger constellation that was called Argo Navis. Argo Navis is Latin for "the ship." Argo Navis includes the star called Puppis (meaning "stern") and the star called Vela (meaning "sails").

Eta Carinae lies within the Carina Nebula and is currently the largest star cluster that can be studied in detail because of its location and size. A nebula is a huge dust

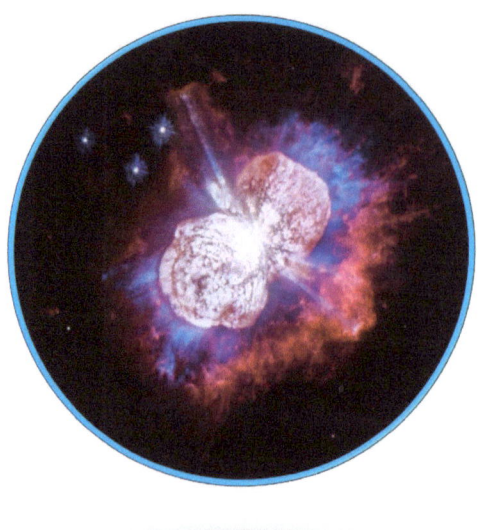

and gas cloud out in space. Some nebulas are formed when a dying star explodes, and, conversely, some are formed when new stars are developing.

A popular NASA telescope image of Eta Carinae is shown above.

Eta Carinae cannot be seen north of latitude 30°N. It never sets below the horizon south of latitude 30°S. Eta Carinae is best seen during the month of March, the month of the spring equinox. Later, you will see why this is important.

In human anatomy, the carina of the trachea is a ridge at the base of the windpipe separating the right and left bronchial tubes. It triggers the cough reflex and has an influence over a person's ability to speak (communicate).

The word Eta in Sanskrit is a girl's name that means "luminous." Sanskrit, which means "perfected" or "refined," is believed to be one of the oldest (if not *the* oldest) human language(s), dating back to the 2nd millennium BC. It is the liturgical language of Hinduism and Buddhism. Sanskrit was considered to be a sophisticated way of speaking and an indicator of high status and education.

Why did our process lead us to this place? Ship, ship's sail, ship's keel, wormholes, luminous, voice (communicate), star destruction, and star creation . . . There is a lot of relevant information contained within this clue. Russian astronauts (and others) have claimed they've seen an angel, or even multiple angels, within this nebula.

6. Eta Islands

There are 3 (and only 3) islands on Earth called Eta Island:

1. Eta Island Canada
2. Eta Island Antarctica
3. Eta Island Bermuda

Taken independently, there is nothing conspicuously special about these places with respect to being centers of ancient cultures, or having the existence of megalithic structures pointing to celestial bodies—at least as far as we know. Perhaps we should mount some expeditions to these places and look more closely. If we look at them together, however, we start to see some very meaningful correlations to the path on which we are traveling in this book. If you draw a line connecting these islands on a map, it looks like this (outside looking in):

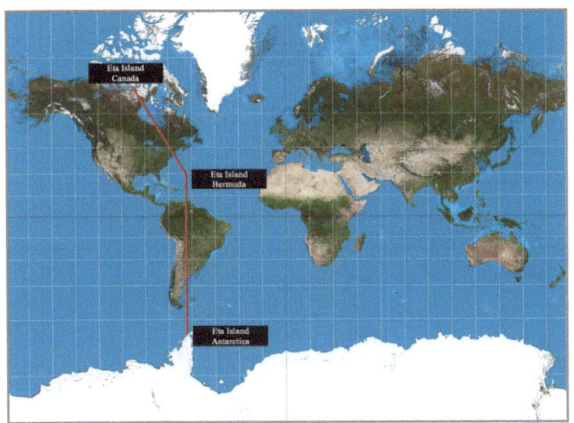

Eta Bermuda and Eta Antarctica are within about 2 degrees of being on the exact same line of longitude. One of the Earth phenomena related to the number 3 is the 3 stars of Orion's belt and their alignment with ancient megalithic structures. There are several ancient megalithic structures that align with the stars of Orion's belt. The most famous are the pyramids of Egypt, shown below. This is the "inside looking out" view of the alignments.

From this angle and distance, the belt looks like a fairly straight line. The line we drew connecting the Eta Islands is not a straight line from that angle and distance. If we draw those lines onto a more three-dimensional model, such as a Google Earth image, it looks like this:

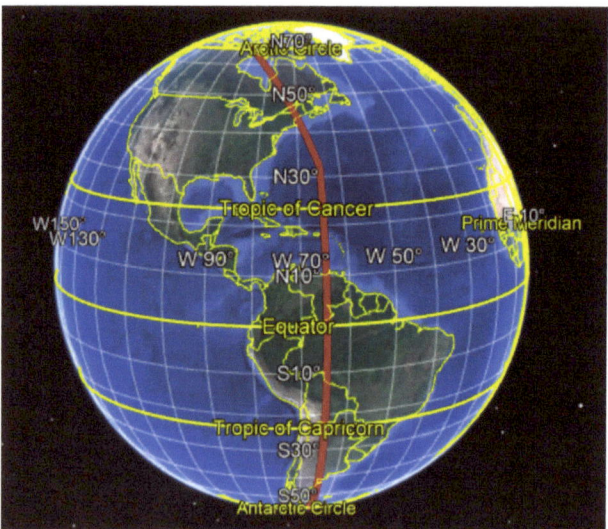

Laying it out on a sphere makes one realize that Orion's belt only looks like a straight line—or close to it—because of the viewing angle and because it's light-years away.

Below is a snapshot taken from a video of Orion shown on the Hubble website. The video goes on to show Orion as the camera passes around it. The video can be found online at https://hubblesite.org/video/5-the-true-shape-of-orion.

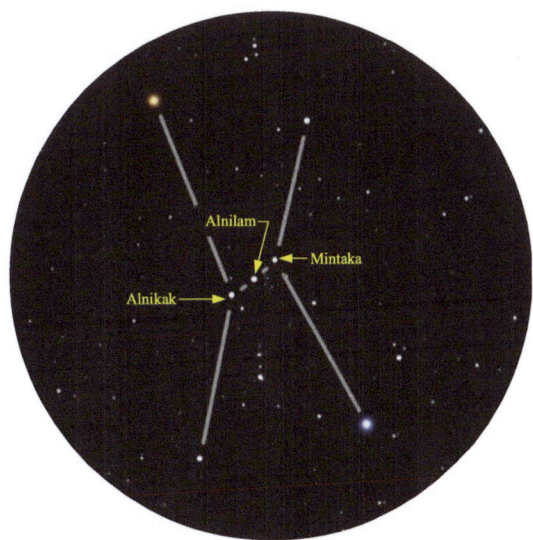

From this angle and distance, it looks like the alignment over the pyramids. This is how we always see Orion in the night sky – a fairly straight line. Notice the hourglass shape previously discussed.

Forty-six seconds into the Hubble video, however, the alignment looks like this:

Then fifty-three seconds into the video:

Rotate the fifty-three-second snapshot ninety degrees clockwise, and it looks a lot like our Eta Island line.

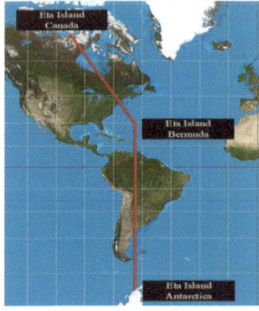 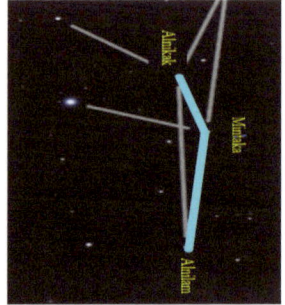

Now compare these alignments to a (Google Earth) view from directly above the pyramids in a position perpendicular to the Earth.

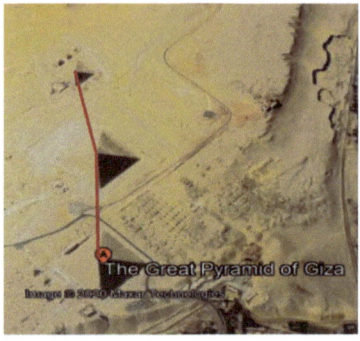

This is the "outside looking in" perspective. This relationship of alignments is also true of other ancient megalithic structures from different cultures on the other side of the world. Now there is an indisputable link between the Orion alignment and what we are doing here—the trail leading to the Eta Island alignment.

The Hubble video goes on to get dynamic footage for approximately two minutes and twenty-three seconds. The point of showing the snapshots from the Hubble video, the Google Earth image alignment, and the Eta Island alignment together is to demonstrate that there are locations somewhere in the universe where an ET would see the belt of Orion in exactly the same alignment as the three Eta Islands. There are many points in space where this alignment would occur as the Earth rotates on its axis and orbits the sun. Perhaps those "outside looking in" locations should be our target search locations and not Orion itself. This concept has never been tested.

Is it possible that when Orion's belt and the Eta Island alignment are in some sort of synchronized alignment together, something happens that opens a gateway for interstellar travel to and from Earth? Since the distances are measured in light-years, there has to be some sort of wormhole or similar portal that allows this to happen. The gap in space and time must somehow be closed in order to make the journey possible, and at this point, we have every reason to believe the journey has been and continues to be made.

NASA published an article in 2012 describing the results of research on "portals" that they funded at the University of Iowa. A portion of that article is shown below.

A favorite theme of science fiction is "the portal"—an extraordinary opening in space or time that connects travelers to distant realms. A good portal is a shortcut, a guide, a door into the unknown. If only they actually existed...

It turns out that they do, sort of, and a NASA-funded researcher at the University of Iowa has figured out how to find them.

"We call them X-points or electron diffusion regions," explains plasma physicist Jack Scudder of the University of Iowa. "They're places where the magnetic field of Earth connects to the magnetic field of the Sun, creating an uninterrupted path leading from our own planet to the sun's atmosphere 93 million miles away."

Observations by NASA's THEMIS spacecraft and Europe's Cluster probes suggest that these magnetic portals open and close dozens of times each day. They're typically located a few tens of thousands of kilometers from Earth where the geomagnetic field meets the onrushing solar wind. Most portals are small and short-lived; others are yawning, vast, and sustained. Tons of energetic particles can flow through the openings, heating Earth's upper atmosphere, sparking geomagnetic storms, and igniting bright polar auroras.

NASA is planning a mission called "MMS," short for Magnetospheric Multiscale Mission, due to launch in 2014, to study the phenomenon. Bristling with energetic particle detectors and magnetic sensors, the four spacecraft of MMS will spread out in Earth's magnetosphere and surround the portals to observe how they work.

To the best of our knowledge, the MMS mission made many interesting discoveries that support the idea of a magnetic influence; however, it did not specifically discover a "portal."

So, what about Orion's belt? One could say this about the synchronization of Orion's alignment with pretty much any three points on Earth. The difference is that we were directed to three specific and unique points through a process based on the binary clues provided. There is a set of directions with clear meaning as to what we are being guided to find. In addition, the Eta Island alignment is a natural alignment, as far as we know. It is not an alignment that was constructed like the pyramids. It seems highly unlikely that the results we are seeing so far are arbitrary, coincidental, or without intelligent design. The significance of Orion's belt is clear and documented by several early and very different cultures; this has to mean something in relation to what we are doing here. There is also the geographic link to factor in.

The next important features of the Eta lines are the end points that lie on two Earth circles. Now we begin to introduce one of the geometric keys to our code: the circle. The Canada Eta Island end is only a couple of degrees north of the Arctic Circle, and the Antarctic Eta Island end is only a couple of degrees north of the Antarctic Circle. One degree of latitude is equal to approximately sixty-nine miles. On a global scale, and even more so on a universal scale, two Earth degrees is nothing. For all intents and purposes, the end points are riding on the Arctic and Antarctic Circles. What is so special about these Earth circles?

The Arctic Circle delineates the northernmost point in the northern hemisphere, where the midday sun is just visible on the horizon on the day of the Winter Solstice. The Winter Solstice is the one day of the year that has the shortest period of daylight. The sun reaches its lowest high point on this day. The North Pole is tilted to its maximum distance away from the sun at this time.

The Arctic Circle also delineates the southernmost point in the northern hemisphere when the sun reaches its highest point in the sky on the day of the Summer Solstice. The Summer Solstice is essentially the opposite of the Winter Solstice. It is the day when the North Pole is at its closest tilt toward the sun.

The Antarctic Circle is the opposite of the Arctic Circle with respect to its relationship to the sun and to daylight. A simplistic diagram of this dynamic is shown below.

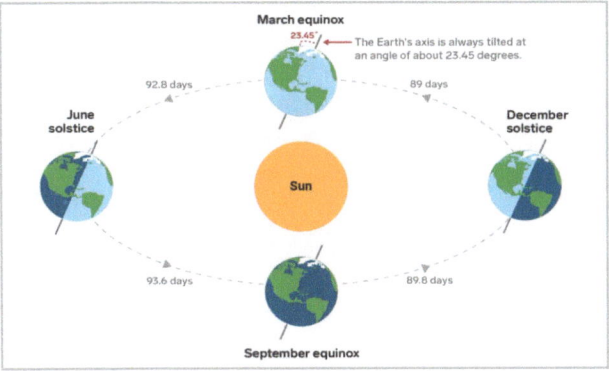

Diagram of solstice source: Timeanddate.com

Notice that the Earth is tilted on its axis by 23.45 degrees or approximately 23.5 degrees. Our first 3 prime numbers (2–3–5) surface again here. The position of the Antarctic and Arctic Circles is not fixed. Their respective latitudes change within a margin of approximately 2 degrees per 41,000 years.

What that means in the context of the significance of the Arctic and Antarctic Circles being the start and end points of the Eta Island lines is this: they are the only two latitudes on Earth where there is exactly one twenty-four-hour period of full daylight and exactly one twenty-four-hour period of twilight. They are unique delineators of night and day. One day of complete light and one day of complete darkness—perhaps an annual reminder of the Alpha and the Omega—the beginning and the end of our Eta lines.

The relationship of earth circles, equinox events and solstice events have been critical to the development of celestial navigation equations and tables for hundreds of years. It is complicated and I am not an astronomer so I offer the following explanation take from:

Bowditch's American Practical Navigator on website: *https://www.starpath.com/celnavbook/BCN_all.pdf*

> *Since the earth travels faster when nearest the sun, the northern hemisphere (astronomical) winter is shorter than its summer by about seven days. Everywhere between the parallels of about 23°26′N and about 23°26′S the sun is directly overhead at some time during the year. Except at the extremes, this occurs twice: once as the sun appears to move northward, and the second time as it moves southward. This is the torrid*

zone. The northern limit is the Tropic of Cancer, and the southern limit's the Tropic of Capricorn. These names come from the constellations which the sun entered at the solstices when the names were first applied more than 2,000 years ago. Today, the sun is in the next constellation toward the west because of precession of the equinoxes. The parallels about 23°26' from the poles, marking the approximate limits of the circumpolar sun, are called polar circles, the one in the Northern Hemisphere being the Arctic Circle and the one in the Southern Hemisphere the Antarctic Circle. The areas inside the polar circles are the north and south frigid zones. The regions between the frigid zones and the torrid zones are the north and south temperate zones. The expression "vernal equinox" and associated expressions are applied both to the times and points of occurrence of the various phenomena. Navigationally, the vernal equinox is sometimes called the first point of Aries (symbol) because, when the name was given, the sun entered the constellation Aries, the ram, at this time. This point is of interest to navigators because it is the origin for measuring sidereal hour angle. The expressions March equinox, June solstice, September equinox, and December solstice are occasionally applied as appropriate, because the more common names are associated with the seasons in the Northern Hemisphere and are six months out of step for the Southern Hemisphere. The axis of the earth is undergoing a precessional motion similar to that of a top spinning with its axis tilted. In about 25,800 years the axis completes a cycle and returns to the position from which it started. Since the celestial equator is 90° from the celestial poles, it too is moving. The result is a slow westward movement of the equinoxes and solstices, which has already carried them about 30°, or one constellation, along the ecliptic from the positions they occupied when named more than 2,000 years ago. Since sidereal hour angle is measured from the vernal equinox, and declination from the celestial equator, the coordinates of celestial bodies would be changing even if the bodies themselves were stationary. This westward

> motion of the equinoxes along the ecliptic is called precession of the equinoxes. The total amount, called general precession, is about 50.27 seconds per year (in 1975). It may be considered divided into two components: precession in right ascension (about 46.10 seconds per year) measured along the celestial equator, and precession in declination (about 20.04" per year) measured perpendicular to the celestial equator. The annual change in the coordinates of any given star, due to precession alone, depends upon its position. Sunlight in summer and winter. Compare the surface covered by the same amount of sunlight on the two dates. Due to precession of the equinoxes, the celestial poles are slowly describing circles in the sky. The north celestial pole is moving closer to Polaris, which it will pass at a distance of approximately 28 minutes about the year 2102. Following this, the polar distance will increase, and eventually other stars, in their turn, will become the Pole Star. The precession of the earth's axis is the result of gravitational forces exerted principally by the sun and moon on the earth's equatorial bulge. The spinning earth responds to these forces in the manner of a gyroscope. Regression of the nodes introduces certain irregularities known as nutation in the precessional motion.

The point in all of this is to understand that mariners throughout history who navigated the globe using the stars; recognized that celestial bodies are predictably moving targets. Adjustments needed to be made to calculations and tables to account for these movements. That is just on an earthly scale. Think about what that means to an extraterrestrial navigator traveling through the universe. Keep this in mind as we move forward.

Now consider the connection point of the Eta Island lines—Eta Island in Bermuda. Why Bermuda?

Before we enter into the dark and vast hall of unsolved mysteries about the Bermuda Triangle, let's look at the island from a geological perspective.

In May 2019, *National Geographic* published an article by Robin George Andrews entitled, "The volcano that built Bermuda is unlike any other on Earth." Without getting into a lengthy dissection of that article, the opening introduction is this:

No two volcanoes are the same, but they all form in the same handful of ways. All, it seems, except for the ancient volcano forming the foundations of the island of Bermuda. After examining rocks from deep under the island, scientists discovered that this quiet volcano formed in a way that is, so far, completely unique. The work, reported this week in the journal Nature, not only solves a long-standing mystery about this beautiful isle in the Atlantic, it also describes a whole new way to make a volcano.

For the sake of argument, we will accept the findings published by a reputable scientific journal. At a minimum, we can say that there is some level of uniqueness in the formation of the Bermuda Islands versus other volcanic islands. There may be a purely geological reason why ETs put Eta Island Bermuda in play as a piece of the puzzle.

Then there is the Bermuda Triangle. Recall that equilateral triangles were previously identified as one of the basic clues for us to use in creating the "map" for ET communication. The Bermuda Triangle is an equilateral triangle. If we put it on the (Google Earth) map along with the Eta Island lines, it looks like this:

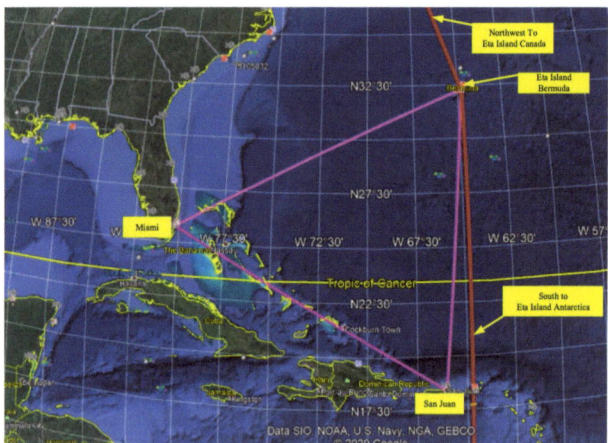

We can see the common point of the triangle connecting to the Eta line at Eta Island Bermuda. We can also see that the southern point of the triangle, San Juan, is only a couple of degrees off the Eta Island longitude line headed south toward

Eta Island Antarctica. We also understand that a couple of degrees on Earth is not much, and it is consistent with the margin of error discussed previously in relation to the location of the Arctic and Antarctic Circles. Consider that our delineation of the Bermuda triangle may be off by a couple of degrees. The northern tip of the Bermuda Triangle is also the approximate northernmost latitude where Eta Carina can be seen in the northern hemisphere, as previously discussed.

The process we are following uses the number 3 as a driver, the first 3 prime

It is reasonable to expect that ETs developing a map for us ...

numbers as the boundaries, and binary code as the language, and it has taken us this far. Now we are introducing the elements of basic geometry. We know that the Earth is a dynamic environment and that the location of geological structures on Earth are not completely fixed forever. There has to be a reasonable margin of error allowed when undergoing any analysis of the location of things on Earth. What is in one spot today was not in that exact same spot forty million years ago. Precise locations of geological markers may have changed over time, but they are close enough for practical use within the time window we are talking about. The further back in time we go with respect to the location of something on Earth, the further away it will be from today's location.

With that in mind, it is reasonable to expect that ETs developing a map for us to find would have set it up such that the eastern leg of their triangle would have fallen directly on top of the Eta line between Eta Island Bermuda and Eta Island Antarctica. After all, we arbitrarily decided the boundaries of the Bermuda Triangle based on actual events in our time in history and nothing else. We also know that things on Earth move over time. Where was Eta Island Bermuda 9,000 years ago in the time of Atlantis? A small adjustment is reasonable.

To put any adjustments to any location into a quantifiable perspective on an Earthly scale, consider that the Earth's circumference is approximately 24,901 miles. One degree of latitude is approximately sixty-nine miles, or approximately 0.28 percent of the Earth's circumference, which equals less than one half of one percent.

On an Earthly scale, a few degrees of adjustment to the location of significant points translates into only a small percentage of the whole.

If we grab the triangle, hold the Eta Island Bermuda point and rotate it to make the eastern leg of the triangle align with the Eta line between Eta Bermuda and Eta Antarctica, it looks something like this:

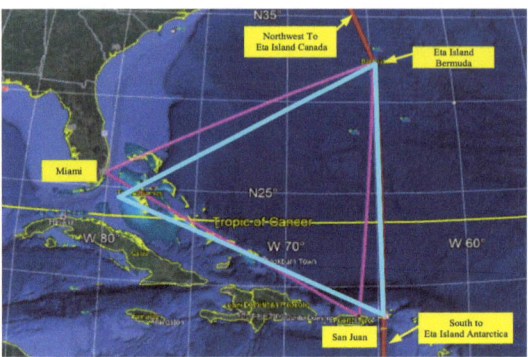

If we close the Eta Island lines and make another triangle by drawing the line that connects the Canadian Eta Island at the Arctic Circle and the Antarctic Eta Island at the Antarctic Circle, it looks like this from outer space:

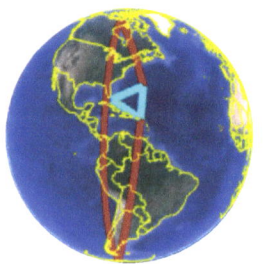

The closer you get, the more the adjustments of alignment show.

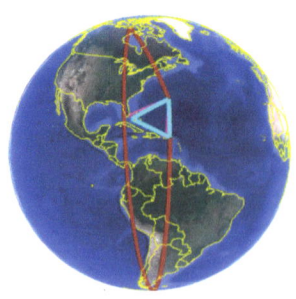

Up close, the rotated Bermuda triangle and its proximity to the western leg of the Eta Triangle looks something like this:

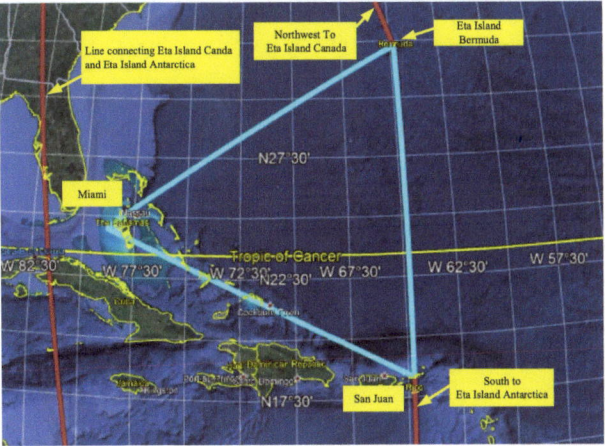

It would make some sense that if ETs were developing some sort of map for us to find using control points to get a fix on position, they would have made the western point of the Eta Triangle lock down on the Eta Island triangle closing line that we show as passing near that point of the triangle. If we extrapolate the rotated triangle to achieve that closure and put all three versions of the triangle together, it looks like this:

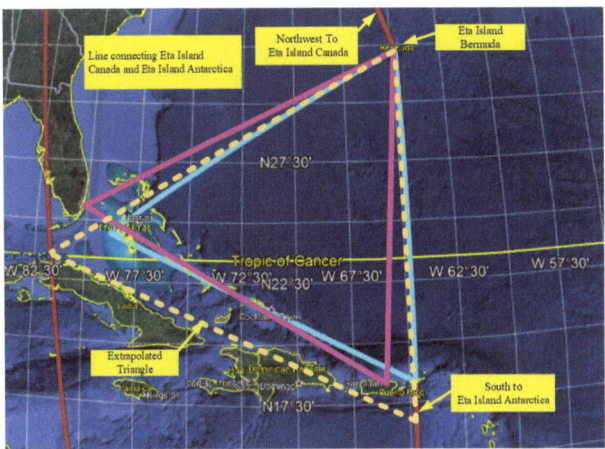

There is not much difference between the three interpretations of the triangle with respect to size and orientation on Earth. Now the Tropic of Cancer, the western point of the "extrapolated" Bermuda Triangle, and the western leg of the Eta Triangle all converge on a single point. The Tropic of Cancer is the northernmost point on Earth where the sun can be observed directly overhead.

This means that within a reasonable margin of error of less than 3 percent in any direction, the Eta Triangle is locked down by three points on these three parallel Earth lines:

1. Arctic Circle—Eta Island Canada
2. Antarctic Circle—Eta Island Antarctica
3. Tropic of Cancer—western point of the Eta Triangle, which also falls on the line connecting Eta Canada and Eta Antarctica.

It should be noted here that the western point of the triangle lies very near the location of the sunken city of Cuba that some theorize to be the location of the lost city of Atlantis. The site lies submerged west of Havana, just off the northwest tip of Cuba's Guanahacabibes Peninsula in the Pinar del Río Province. The depth of the site is between 2,000 and 2,460 feet, and it covers approximately 0.8 square miles.

The site was first recorded using sonar imagery in 2001 by a Canadian company (Advanced Digital Communications). They returned to the site a second time and took pictures using a submersible Remotely Operated Vehicle (ROV).

The images are not clear enough to be definitive, and at that depth, it is difficult and costly to reach the site with the technology necessary to obtain better images. The original sonar images are shown below on the left compared to a computer-generated image using the sonar data on the right.

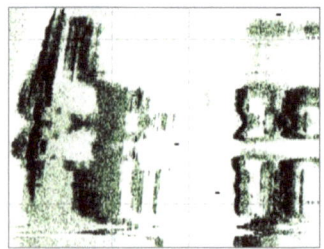 

The images have been interpreted by some scientists as man-made structures and by others as naturally occurring. The truth remains a mystery to this day, but the possibility exists that they are man-made.

The Eta Triangle's western point lies within approximately 3 degrees of longitude and approximately 2 degrees of latitude of the sunken city's location. That is fairly close on a universal or even global scale.

Triangulation is an ancient, tried-and-true method for a ship to determine or "fix" its position by taking bearings from three fixed objects on the ground and/or celestial bodies (or satellites). Is it possible that this is what we are looking at here with the Eta Triangle?

Much has been written about the Bermuda Triangle and all the unexplained phenomena that have happened within its boundaries. We are not trying to reopen that discussion here, although it seems that there may be a link to this place on Earth (and those phenomena) and the results of our analysis so far. What is significant is that humans defined the boundaries of the triangle semi-arbitrarily, based on the frequency and nature of the many unexplained phenomena that have occurred within it. We, however, were brought to this location without considering any of those phenomena. We were brought to this location by the clues yielded from the binary model that created the trail for us to find and follow. The same result was obtained via two independent processes carried out by unconnected methods. Recall that this "independent verification" was also the case when we examined the International Morse Code earlier. The same result was obtained via two different methods carried out by two independent parties. That is fairly strong evidence that there is merit to our methodology, and it has to mean something. At the very least, it lends more credibility than has ever been offered in the past to some of the celestial and ET-based explanations of the Bermuda Triangle mysteries and gives the skeptics something to consider.

As interesting as all this may be, our findings so far are nothing more than indicators that we are on the right trail to some bigger answers. We need to stay focused and on course to the end.

The three Eta Islands and the three Earth-specific latitude circles have essentially defined our triangle of interest. Now, let that triangle define a circle of interest to narrow down our focus area. For our purposes, we will continue to use the "rotated and extrapolated" triangle. It will become obvious that because of scale, whichever triangle is used does not significantly impact the eventual findings. One of the unique properties of the equilateral triangle is that its centroid and the center of the inscribed circle that it can hold are the same point. An inscribed circle within the triangle looks something like this (note the location of the sunken city):

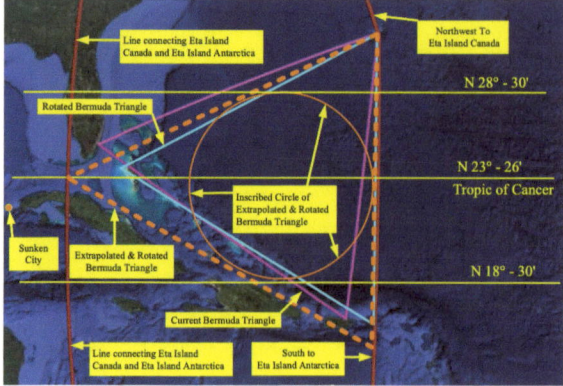

If we simplify all of these lines to focus on the extrapolated triangle and the inscribed circle, it looks like this:

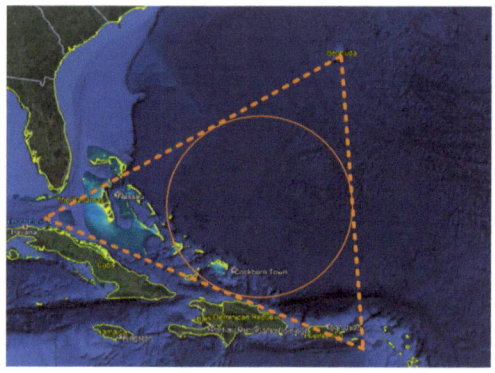

It looks conspicuously like a simplistic and mathematical (not artistic) version of the famous "all-seeing eye" that appears in so many important places throughout the history of humanity.

The all-seeing eye is the subject of much discussion regarding ET influences on Earth. It also looks a lot like the underbelly of a UFO as described by many UFO eyewitnesses. Is it possible that this circle we were led to is a beacon for us to point to the stars (inside-out view) and/or vice versa—for ETs to find their way back here via a portal (outside-in)? The circle is shown below, bounded by Earth latitudes and with the Tropic of Cancer shown.

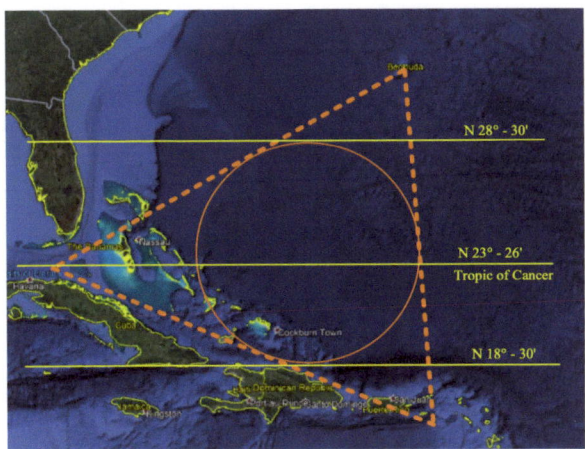

The distance between the upper latitude and the lower latitude is approximately 10 degrees or 690 miles. Note that the sum of our 2–3–5 binary code boundary condition is also (2+3+5) 10. This represents less than 3 percent of the Earth's circumference, so it's a small target. Also, note that the Tropic of Cancer lies approximately at latitude 23° 26'. Twenty-six minutes is approximately 30 minutes (within a small percentage of error) or 0.5°. Therefore, the latitude of the Tropic of Cancer is at 23.5°N (2–3–5), which runs through the exact center of the Eta circle. Recall that 23.5 spins per second is also the rotation of Radio Pulsar PSR J0250+5854. It is also the degree of tilt of the Earth's axis.

Think about this for a minute.

a. 2–3–5 corresponded to CEQ in our first search for letters in the binary block. As previously discussed, this is pronounced "seek you." Eliminate the "E" that corresponds to the number 3 from our analysis, and you are left with the letters CQ, which is also phonetically pronounced "seek you." The call-out letters "CQ" on amateur/HAM radio are used when the radio operator is asking anybody out there who is listening to respond.

b. 2–3–5 is a feature that describes the spin speed of the slowest known radio pulsar. We included pulsar information on our Voyager message in the 1970s.

c. 2–3–5 is a feature of the Earth and its axial tilt angle.

d. 2–3–5 is the latitude of the Tropic of Cancer, which splits our Eta circle right down the middle.

e. 2–3–5 adds up to 10, which is also the number of degrees of latitude bounding the Eta circle.

f. 2–3–5 is the foundation of the binary block that lead us to these clues that describe communication, speed, angle, distance, and location.

This whole process started with the first three prime numbers, 2, 3, and 5. The circle and lines are not presented with pinpoint accuracy here, but they are presented with a very small and reasonable percentage of error. We are slowly and methodically defining a fairly specific area of significance with which to continue with this process. Moving to a global, and then to a universal scale will render any location-based discrepancies at this close-up scale irrelevant.

We have followed a long trail of mathematically and geographically derived clues to get to an Eta-derived circle of this particular size and in this particular location. Is it possible that this is the main key we were meant to find? If it is, we next need to figure out how to use it.

It seems that there are two ways to use this tool. The first is an "inside looking out" approach. Think of the Eta circle as a spotlight or beacon pointing out into the universe on a fixed rotation. We can lock its position on the Earth down right where we found it (on control points and lines) and let it rotate around with the Earth on its 23.5° axis. It would be a pinpoint laser if that radial shaft of light (or radio signal) were projected out into the universe.

If we tracked that projection over one full revolution of the Earth on its axis (one day), the beam would travel in a circle, and we could look at what the beam hits as it travels around that circle. Perhaps it lands on specific targets of interest at particular times . . . But which times? The answer is probably in the clues already provided. Given that our search has found clues linked to the latitudes where there are important relationships on this Earth defined by the sun, it is reasonable to suspect that maybe the days we should focus on are:

1. The vernal equinox in March
2. The summer solstice in June
3. The autumnal equinox in September
4. The winter solstice in December

If we can somehow match this up with the locations in the universe where the view of Orion's belt from outer space is in some sort of alignment with our Eta Island line alignment and determine when that alignment occurs, we will probably be getting very close to answering the questions of where ETs came from and where they go. Maybe this beacon shines on the location of the outer space end of the portal that connects to the Eta circle Earth end of the portal.

Our efforts to date have focused on searching the sky randomly and trying to capture a uniqueness in the twinkle of a light in the sky or a new noise and theorizing about what that twinkle or noise might mean. Of course, there is much more to it than that—no disrespect intended to the extremely brilliant astrophysics community in any way. The point is that maybe it's time to introduce a fresh approach to the search to get better results regarding the establishment of ET communication. This fresh, new approach should probably also consider the Earth's precession.

Precession is the fluctuating axial tilt of the Earth: the slow, but constant gyration of the Earth's axis. It takes approximately 26,000 years to complete one full gyration. This Earth dynamic was first discovered by Hipparchus in the second century BC. The simplest explanation is that the Earth moves like a spinning top, so receiving and sending messages to and from ETs involves a moving target. A diagram of this dynamic is shown below.

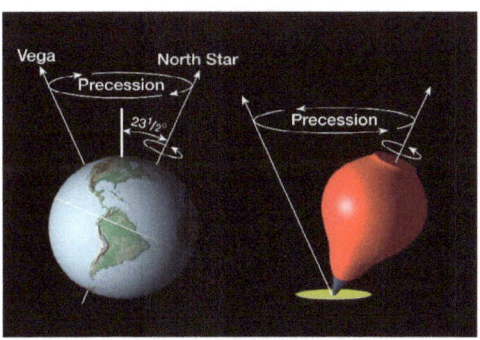

The possibilities certainly get one thinking differently. This dynamic may be involved in the identification of the direction in which to look out into the universe and narrow the search. It may help identify the portal that ETs use, which we found using ET-provided clues. This is a very different approach to the problem.

The second way to use the Eta circle tool is an "outside looking in" approach. What happens if we lock down the Eta circle within the Eta Triangle and let the Earth rotate *under* it versus rotating *with* it? If we did that, it would track the shadow of the circle around the globe guided by the Eta triangle's controlling latitudes of the Arctic and Antarctic Circles, plus the Tropic of Cancer. We would see where on Earth this circle may have influenced humanity and perhaps where portals may historically have opened. We have to remember that ETs probably are (or were) not dialed into our man-made longitude and latitude lines. They probably used geological features and the sun as guides. Think of the Arctic Circle, the Antarctic Circle, and the Tropic of Cancer

What goes on within these ten degrees of latitude around the globe?

as tracks on which our Eta Triangle and its inscribed circle could roll. The Eta circle would traverse the globe, and we could see what it passes over. Putting the Eta circle on parallel circle tracks would keep the Eta circle locked between its upper and lower latitude lines. What goes on within these ten degrees of latitude around the globe?

The yellow band across the Earth map shown below represents the locations the Eta circle would pass over as the Earth rotated under it on its axis or the upper and lower latitudes that limit the circle. The numbered locations highlight some ancient structures of significance that lie within this band of latitude. The point is to demonstrate that the band goes right through the Mayan civilization and the Egyptian civilization. It also covers the Mexico Zone of Silence (item 1 on the map). These places hold some of the most significant ancient civilizations and structures currently known to humanity. The width of this band is a small percentage of the Earth's latitudinal circumference. There are a lot of significant ancient sites in a relatively small area.

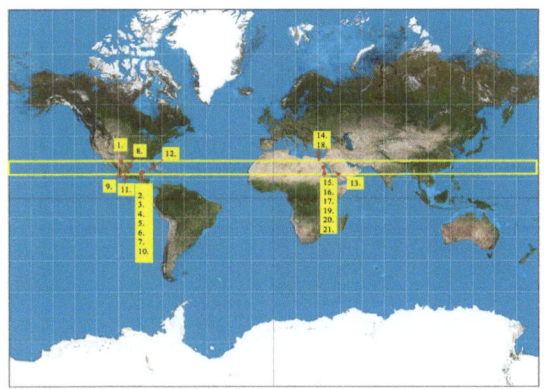

The numbered locations on the map correspond to the following ancient civilization sites:

1. Mexico Zone of Silence
2. Chichen Iza
3. Tulum
4. Ek Balam
5. Coba
6. Uxmal
7. El Rey
8. El Tajin
9. Tula
10. Labna
11. Templo Mayor
12. Cuba Sunken City
13. Mecca
14. Pyramids of Giza
15. Temple of Karnak
16. Abu Simbel Temple Complex
17. Valley of the Kings
18. Saqqara Pyramids
19. Luxor Temple
20. Temple of Hatshepsut at Deir el Bahari
21. Midinet Habu

Several of these ancient locations have structures that align with either Orion's belt, or solstice/equinox events. The Mexico Zone of Silence has been called the Bermuda Triangle of Mexico because there is a history of the same sort of unexplained phenomena that have occurred there with respect to radio signal interference and UFO sightings. It is dry land now, but it was under the ocean millions of years ago.

The Zone of Silence is located between latitude 26 and latitude 28 in the Chihuahuan Desert near Durango, Mexico. It was first identified as an area of interest in the 1930s by pilot Francisco Sarabia. He reported that his radio stopped working, and his plane's instruments went out of control while flying over this area. Scientists figured out that local magnetic fields prevent radio signals from being transmitted in this area. The area was given the name "Zone of Silence" by an oil company in 1966. The Pemex oil company sent out a crew of people looking for promising locations for oil drilling, and they experienced difficulty with radio transmissions/communications, and so the "Zone of Silence" was named.

The area is known to be a regular target for meteors falling to Earth. It also holds several unique animal species and bizarre mutations of plants and animals. In addition, there have been numerous reports of ET and UFO sightings, including direct ET contact with humans, in this zone.

The reported phenomena were sufficient for the Mexican government to establish the Mapimí Biosphere Reserve within the Zone of Silence. The official purpose of the facility is to protect and study the unusual plant and animal life of the area.

Many believe that there is a lot more going on there with respect to what is being researched. It seems unlikely that the Mexican government created such a vast bio-protection zone in the middle of a "silent zone" in the desert for the sole purpose of studying the handful of unique plants and animals that live there.

The observation that the Zone of Silence, the Mayan Pyramids, the Bermuda Triangle, and the Egyptian Pyramids fall within the same small latitudinal zone of Earth has been made by others, as shown previously on the "Eta Band" world map. In addition, much has been written about the numerous correlations between the Mayan and Egyptian civilizations and their respective megalithic structures. What is important here is that we are now driven to this observation by way of constructing and then interpreting a block of binary code we were led to by clues that may have been provided by ETs—specifically, the number 3. Now this alignment and cultural similarity between phenomena have a mathematical link beyond just our Earthly observations: independent verification for the third time so far.

If we put the Eta band on a global map with a focus on world religions, it looks like this:

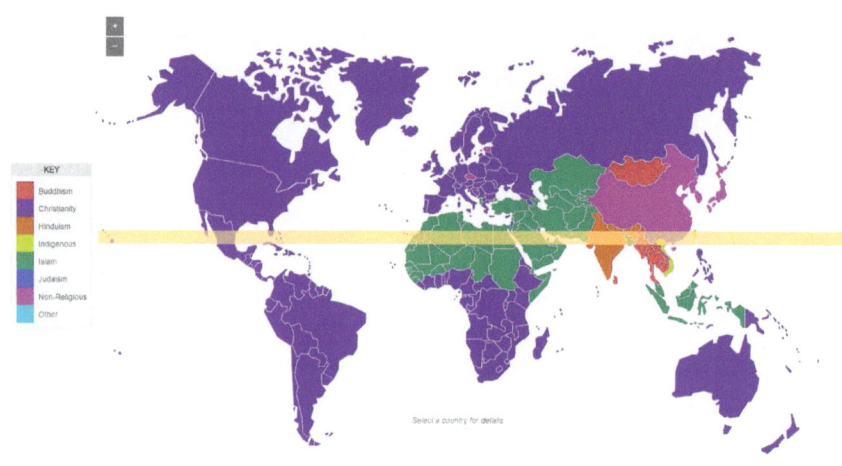

The map is taken from website http://d3tt741pwxqwmo.cloudfront.net/WGBH/sj14/sj14-int-religmap/index.html#: and it color codes the major world religions based on the beliefs of the majority of the population in each region. There is a lot of relevant information on this map as it applies to the Eta band.

Examination of the map shows that within a reasonable margin of error:

1. The Eta band splits the geographical area where Christianity—the most popular religion—is most prevalent more or less in half north to south; however, it touches very little of the Christian landmass.

2. The Eta band splits the Islamic geographical area more or less in half north to south, and it also has a very large contact area within the most populous Islamic regions.

3. The Eta band splits the geographical area covered by the third most popular religion, Hinduism, more or less in half north to south.

4. The Eta band barely touches the southernmost point of the nonreligious area, and it splits the area of Buddhism north to south.

5. Christianity is the religion that has the largest number of adherents across the globe.

What does this all mean in the big picture? The Eta circle essentially passes over and splits the regions holding the top 3 religions, with followers of Islam more or less in the middle. Again, the number 3 surfaces.

It appears so far that the word "Eta" generated by our analysis is sending a message about location and communication—the location of important and relevant things on Earth, and the location of potential search targets in the universe. It also suggests the location of the Earth end of a wormhole that we need to access in order to establish communication.

There is more we could discuss about the importance and meaning of the word "Eta," but the point is made. It seems very possible now, based on some evidence, that the Eta circle might become a wormhole-like portal at specific times of the year when the alignment of Orion's belt is just right. We were guided to Eta through a process. It was the first word revealed in that process, and clearly, it has a lot of meaning related to the premise of this book.

Moving on from Eta and recalling our original list of words that were derived from our process, those words were:

2 = ETA
3 = PAS
5 = SWY
7 = ZAC

PAS is the second word our process revealed, corresponding to prime number 3, which is the most important number used in this analysis. Researching the word PAS does not lead down as many trails as Eta, but it does say something important.

In French, Pas is a step or series of steps in a ballet; a track or passage. Like Eta, PAS also leads us to a "track" in the way we described the Eta circle as being on a track of three parallel Earth circles.

In microbiology, there are PAS proteins that were recently discovered. The American Society for Microbiology published an article on the subject in 1999 (*Microbiology and Molecular Biology Reviews*, June 1999, 63(2): 479–506). Here is the abstract from that article:

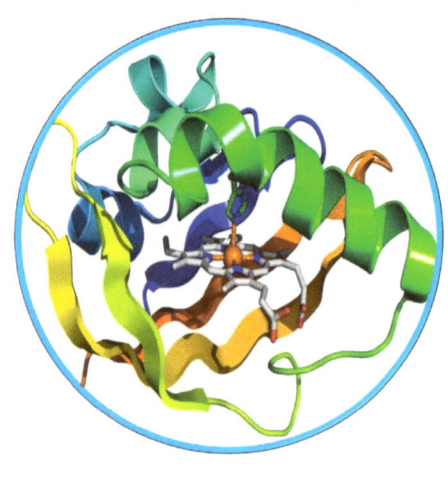

> *PAS domains are newly recognized signaling domains that are widely distributed in proteins from members of the Archaea and Bacteria and from fungi, plants, insects, and vertebrates. They function as input modules in proteins that sense oxygen, redox potential, light, and some other stimuli. Specificity in sensing arises, in part, from different cofactors that may be associated with the PAS fold. Transduction of redox signals may be a common mechanistic theme in many different PAS domains. PAS proteins are always located intracellularly but may monitor the external as well as the internal environment. One way in which prokaryotic PAS proteins sense the environment is by detecting changes in the electron transport system. This serves as an early warning system for any reduction in cellular energy levels. Human PAS proteins include hypoxia-inducible factors and voltage-sensitive ion channels; other PAS proteins are integral components of circadian clocks. Although PAS domains were only recently identified, the signaling functions with which they are associated have long been recognized as fundamental properties of living cells.*

The clue here is about life at its most fundamental level. We did a very similar thing by sending coded instructions about our DNA into the universe for ETs to discover and decode. Perhaps we are different from ETs on an evolutionary scale, but our thought processes may be very similar.

In ancient Hebrew, Pas means: (1) individually—each, every, any, all, the whole, everyone, all things, everything, and (2) collectively—some of all types.

PAS appears to be a simple way to communicate and reference "life" collectively— "some of all types." Communicating with life not of this Earth is the ultimate goal, and our process is now going in that direction. Is it not the most important question for which we seek an answer? Finding intelligent ET life and establishing communications is the goal. Are ETs signaling to us that life is out there? Are they indicating that a "track or passage" to the PAS proteins is out there and that they are "the fundamental properties of living cells— "all types"? That's it—a simple message. Is that not what we are trying to communicate with out there—other forms of life? It is certainly a much simpler way to communicate "life" than the way in which we tried to communicate by sending out the code and chemistry for our double helix DNA.

Swy is the third word revealed in our process, which corresponds to the prime number 5 position on our list. Like Pas, researching Swy does not seem to lead down many trails for clues compared with Eta. Like Pas, however, it does give an important but simple message.

In Hebrew and in Aramaic, there are three interpretations of the verb SWY:

1. Be equal
2. Be like
3. Be swift

Is the message that human beings are similar or equal to the ETs? Are they telling us that we will be swift in contacting them now that we have figured out this message? Maybe it's a warning to hurry up because something bad is in the near future. Pas is life, and perhaps Pas and Swy are the clues telling us that the life out there can be likened to or is equal to ours. The message might just be that basic.

There are numerous Bible references about God remembering someone.

We have already mentioned what Zac, the last word in the list, corresponding to prime number 7, means. It refers to the 7th letter of the Greek alphabet—Eta—which turned out to be an important clue in our analysis. It too originally meant the number 7. In addition, the Bible tells us that on the seventh day, God rested from creating the universe and all the life that it contains. Zac is also an ancient Hebrew boy's name that means "God remembers."

What could that mean in the context of what we are doing and its appearance in this process? Why would ETs stick this at the end of the message? There are numerous Bible references about God remembering someone. God never forgets anything, and saying he "remembers" implies he forgot someone, but that is not the case. "Remember" translates from the ancient Hebrew language into meaning that God focused his attention on someone and acted on their behalf. The ancient Hebrew concept and use of the word "remember" always involves thinking of a person and acting on their behalf in a positive way. Every passage of the Bible that says, "God remembered" is followed by some sort of action or work on behalf of the person or people being remembered. When God "remembered" Noah and his family floating in the ark, he caused the wind to blow, which began to dry up the water that covered the Earth, as cited in Genesis 8:1:

> *But God remembered Noah and all the beasts and all the livestock that were with him in the ark. And God made a wind blow over the earth, and the waters subsided.*

Is the message or clue presented to us from ETs telling us that they "remember" or are thinking about us and are willing to act on our behalf? Are they acting on our behalf to help us demonstrate our state of readiness to enter the community of the universe?

Perhaps we have figured out some important keys to open the door to a solution to our quest. The takeaway summary from our analysis of the model and our interpretation of what it might mean so far is this:

1. ETs probably came to Earth on a ship from the stars by way of a portal.

2. Radio signals are probably the long-distance communication media of choice.

3. The Eta circle is a solid candidate for being the Earth end of the portal.

4. The Earth end of the portal probably moves around the Earth on a fixed track.

5. The portal might only open during solstice and equinox events.

6. Shine the portal out into space, and we might find out where the other end of the portal opens (when it opens).

7. Life forms that are similar or equal to ours probably await at the other end of the portal.

There is no doubt that some people will read this book and immediately present their case that what we have found and how we have found it is inaccurate. It has already been stated that slightly different results have been found right out of the gate. That is fine, because if you throw the entire process that came from the binary block out of the window and pick up the trail using only the lineup of the three Eta Islands and the Bermuda Triangle, the conclusions based on that information alone remain real and relevant. Anybody could have made the Eta Island observation and started down this road from that point; however, being guided to this point by supporting information makes a significant difference to the credibility of my conclusions.

(Upper- and lowercase Greek letter Eta)

7.

Response to the Message

WHAT ARE WE SUPPOSED TO do with all of this information? We are supposed to respond and keep the response to a simple "repeat-back" of what we discovered. Basic communication protocols require that we follow the three-way communication procedure. If the ETs have indeed sent us down a relatively simple trail to the message of Eta, Pas, Swy, and Zac, we should repeat the message back (in binary code), consistent with the procedures for three-way communication. We should send the message we discovered back to them exactly as it was discovered. This will clearly demonstrate that we have figured out the message, understood it, and completed the action required—which is to repeat it back. The contents of the response are important, but the pattern and timing of its transmission is equally important.

It would seem logical to send the response message in patterns of 3 and 7 (the Morse code pattern). There is more than one way to do this, but I will present what I believe is the strongest option.

One option for the message is to focus on the 2–3–5–7 pattern that was the driver of all the discoveries cited in this book. One possible response message that would make sense is shown below, with the understanding that only the binary digits would be transmitted. The letters (in color) that correspond to their binary values are shown in the graphic to demonstrate the pattern.

There are probably more options, and all ideas are encouraged and welcomed. To accomplish the intent of the message, the boundary conditions are these:

1. Our message has to follow a pattern that will best demonstrate that we are *responding* to the ET message as opposed to sending an original message of our own.

2. Our response has to be structured using the prime number patterns that got us this far in the process.

	Midnight on Day Before Equinox / Solstice		Midnight on Day of Equinox / Solstice		Midnight on Day After Equinox / Solstice	
	E	01000101	E	01000101	E	01000101
3 Sec.	T	01010100	3 Sec.	T	01010100	3 Sec.	T	01010100
	a	01100001		a	01100001		a	01100001
3 Sec. Gap			3 Sec. Gap			3 Sec. Gap		
	P	01010000		P	01010000		P	01010000
3 Sec.	a	01100001	3 Sec.	a	01100001	3 Sec.	a	01100001
	S	01010011		S	01010011		S	01010011
3 Sec. Gap			3 Sec. Gap			3 Sec. Gap		
	S	01010011		S	01010011		S	01010011
3 Sec.	W	01010111	3 Sec.	W	01010111	3 Sec.	W	01010111
	y	01111001		y	01111001		y	01111001
3 Sec. Gap			3 Sec. Gap			3 Sec. Gap		
	z	01111010		z	01111010		z	01111010
3 Sec.	a	01100001	3 Sec.	a	01100001	3 Sec.	a	01100001
	C	01000011		C	01000011		C	01000011
7 Sec. Gap			7 Sec. Gap			7 Sec. Gap		
	E	01000101		E	01000101		E	01000101
3 Sec.	T	01010100	3 Sec.	T	01010100	3 Sec.	T	01010100
	a	01100001		a	01100001		a	01100001
3 Sec. Gap			3 Sec. Gap			3 Sec. Gap		
	P	01010000		P	01010000		P	01010000
3 Sec.	a	01100001	3 Sec.	a	01100001	3 Sec.	a	01100001
77 Seconds per Set	S	01010011		S	01010011		S	01010011
3 Sec. Gap			3 Sec. Gap			3 Sec. Gap		
	S	01010011		S	01010011		S	01010011
3 Sec.	W	01010111	3 Sec.	W	01010111	3 Sec.	W	01010111
	y	01111001		y	01111001		y	01111001
3 Sec. Gap			3 Sec. Gap			3 Sec. Gap		
	z	01111010		z	01111010		z	01111010
3 Sec.	a	01100001	3 Sec.	a	01100001	3 Sec.	a	01100001
	C	01000011		C	01000011		C	01000011
7 Sec. Gap			7 Sec. Gap			7 Sec. Gap		
	E	01000101		E	01000101		E	01000101
3 Sec.	T	01010100	3 Sec.	T	01010100	3 Sec.	T	01010100
	a	01100001		a	01100001		a	01100001
3 Sec. Gap			3 Sec. Gap			3 Sec. Gap		
	P	01010000		P	01010000		P	01010000
3 Sec.	a	01100001	3 Sec.	a	01100001	3 Sec.	a	01100001
	S	01010011		S	01010011		S	01010011
3 Sec. Gap			3 Sec. Gap			3 Sec. Gap		
	S	01010011		S	01010011		S	01010011
3 Sec.	W	01010111	3 Sec.	W	01010111	3 Sec.	W	01010111
	y	01111001		y	01111001		y	01111001
3 Sec. Gap			3 Sec. Gap			3 Sec. Gap		
	z	01111010		z	01111010		z	01111010
3 Sec.	a	01100001	3 Sec.	a	01100001	3 Sec.	a	01100001
	C	01000011		C	01000011		C	01000011
3 HOUR GAP			3 HOUR GAP			3 HOUR GAP		
Repeat 2 times = 3 Reps			Repeat 2 times = 3 Reps			Repeat 2 times = 3 Reps		
15 Hours to Next Day Repeat			15 Hours to Next Day Repeat			End Response		

We should launch our message through the portal revealed in the ET message to us—the Eta circle. The question is: in what direction do we send the message? Do we send it out in a direction perpendicular to Earth in the hopes that we are driving it down the throat end of a portal? Or do we send our message from the Eta circle in some direction linked to the alignment of Orion's belt? More research on this needs to be done to increase our chances that our response will be received in a timely manner. Sending a message into space without sending it through a portal seems futile. The distances are too great; it would take hundreds or thousands of years to be received. For all we know, hitting the portal could shrink the time to hours or even minutes. Exchanging messages that take forever to receive is not very helpful in establishing any meaningful communication with ETs from other galaxies or even from our own galaxy.

> They are actually very close, and they might just be waiting for us to deliver the correct form of response.

However, given that we believe that ETs are here fairly often, the time gap might be less of a problem than we realize. They are actually very close, and they might just be waiting for us to deliver the correct form of response. It is not an unreasonable possibility.

If we were going to propose a procedure for receiving and sending messages based on the information revealed in this book, where would we start? If this were a life-or-death situation (which it may very well be) and we only had one bullet in the gun, what would we shoot at? To identify the best target, we should use the information we've uncovered so far.

Thanks to our discovery of the Eta Island lines and all of the other evidence indicating its relevance, we must believe that Orion's belt is involved, so let's start there. Orion is visible from October through March; this period is the only time it can be seen from Earth because of the Earth's orbit around the Sun.

Within the Eta circle, Orion's belt pretty much looks like this:

From Eta Island Canada (the Arctic Circle), it lies further southwest and looks like this:

From Eta Island Antarctica (the Antarctic Circle), it lies further northwest and looks like this:

There are two relevant solar events that occur during the time span when Orion is visible from Earth: the winter solstice in December, and the vernal equinox in March.

Now let's look at some important angular relationships with Orion's belt. We previously saw how Orion's belt vertically aligns with the Great Pyramids of Giza. The angle of that alignment looks like this:

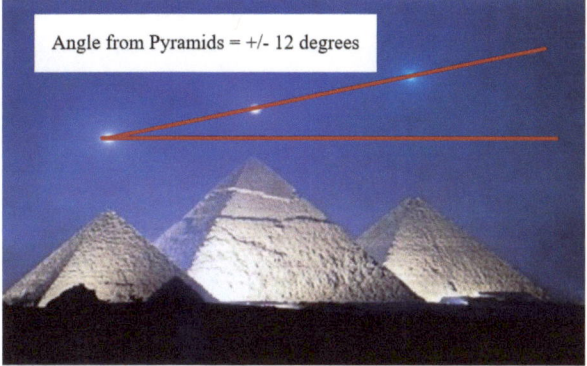

The angle of Orion's belt is approximately 12 degrees from the horizontal. Recall that the Earth's axial tilt is 23.5 degrees. Twelve degrees is almost exactly half of 23.5, bringing our 2–3–5 series into play again.

The following three views of Orion's were taken from an interactive map from globeatnight.org at approximately 08:00 PM in March 2017.

> *The shape of different constellations and asterisms changes over time as a result of stars not being stationary and moving through space, but the three stars of the Orion's Belt share the same origin and have the same proper motion, which means that they travel together, with the asterism retaining a similar shape over the centuries. This means that the Orion's Belt looked almost the same in ancient times as it does now.*
> *—Constellation Guide. June 2014. Orion's Belt.*

From the Arctic Circle, the angle of Orion's belt looks like this:

Angle from the Arctic Circle = +/- 23 Degrees

From this location, the angle from the horizontal is approximately 23 degrees, which is the angle of the Earth's tilt on its axis within a very small margin of error.

From the Antarctic Circle, the angle looks like this:

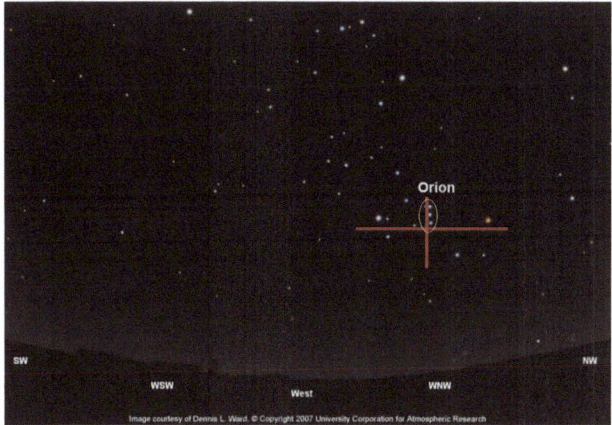

Angle from the Antarctic Circle = +/- 90 Degrees

It is perpendicular to the horizon, which is the direction I suggest we aim our message from the Tropic of Cancer within the Eta circle.

From the Eta circle, the angle looks like this:

Angle from the Tropic of Cancer = +/- 44 Degrees

From this position, the angle is approximately 44 degrees from the horizontal. The difference between the Arctic Circle angle (23 degrees) and the Eta circle Tropic of Cancer angle (44 degrees) is 44 – 23 degrees = 21 degrees. This is, again, the angle of the Earth's axial tilt within a very small margin of error.

If we put all of these angular relationships together on a protractor, it looks like this:

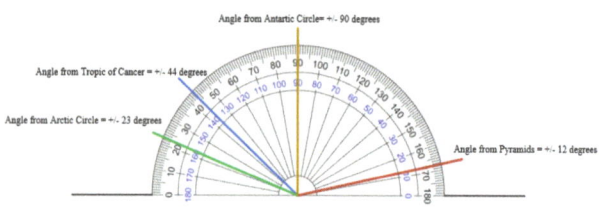

If we subtract the angular values for the Eta circle at the Tropic of Cancer (44 degrees) and the angle from the Arctic Circle (23 degrees) from the Antarctic Circle angle (90 degrees), the result is 90 – 44 – 23 = 23 degrees. Once again, we get the angle of the Earth's axial tilt. We could go one step further and subtract the 12-degree angle created by the Great Pyramids of Giza (12 degrees), and the result would be 23 – 12 degrees = 11 degrees. The resulting 11 degrees is essentially equivalent to the pyramids' angle, which is approximately half of the 23.5-degree Earth's axial tilt angle within a very small margin of error. If we add the angular values, we get 90 + 44 + 23 + 12 = 169 degrees. The supplementary angle to the 12-degree pyramid angle is 169 degrees within a very small margin of error. Supplementary angles are two angles whose sum is 180 degrees (169 + 12 = 181). All of these angular relationships imply a link to our process with Orion's belt.

What these angular relationships further imply is that all of the ancient structures that point at Orion's belt may be an angular reference for navigational purposes versus a reference to the location of the constellation itself. Orion's belt might just be a marker or navigational beacon of some sort. If we think of it this way (differently than we have in the past), all of a sudden, the ancient Orion references start to make some quantifiable, practical, and tangible sense versus what we have now, which is speculation as to what the alignment of Orion's belt might mean with respect to ancient structures.

So, what is the best way to initiate a new ET search and response campaign based on what we have revealed by our process so far? The easiest thing to do might be to position a radio telescope, along with whatever other sophisticated listening devices exist, on a ship and set it up on the Tropic of Cancer in the center of the Eta circle in the Bermuda Triangle location. From that location, we would both listen for a signal, and send out our response. Our receivers and launch should be aimed straight up, as perpendicular to the Earth's surface at that point as possible. This should be done for a couple of days or so around the solstice and equinox events when Orion's belt is visible. The crew on the vessel should include, at a minimum, several people charged with the task of observing the sky through high-powered binoculars, hawking the skies for UFO activity. If the wormhole is indeed there and open, we might observe some sort of ET activity not related to our message response activity, i.e., two birds with one stone, if you will.

The Arecibo radio telescope previously discussed is near the Eta Circle in Puerto Rico and has been listening to the sky for decades with no obvious signal from an ET. There are many possible reasons why no signal has been heard, including that there may be no signal to hear. Maybe they are just missing the wormhole when it is open? They are outside the Eta Circle, so who knows? They know (or at least believe) that the messages they do receive are thousands of years old. If an ET did respond to the 1974 Arecibo message in the same direction from which it came (not through a wormhole), we are many hundreds of years away from receiving it. What we do have now is a direction in which to turn our ear, a timeframe in which to listen hard, and a location from which to transmit and listen. To be successful, the send/receive time gap has to be closed somehow, and the Eta circle is a good candidate for a wormhole location.

Can we use an existing radio telescope to test our theory? There are over 100 large radio telescopes in operation throughout the world. This represents billions of dollars in construction, operation, and maintenance. The SETI (Search for Extra-Terrestrial Intelligence) organization's purpose is to analyze radio signals, searching for signs of extraterrestrial intelligence as one of many activities undertaken as part of a worldwide effort. Their mission is to explore, understand, and explain the origin and nature of life in the universe and the evolution of intelligence. A world map is shown below, taken from IUCAF's (Inter-Union Commission on Frequency Allocation for Radio Astronomy and Space Science) inventory of radio telescope locations found on this website: www.google.com/maps/d/viewer?msa=0&mid=1HX_7mIUcmDybmsONe9hJGisMOhc&ll=-20.364664620722447%2C-2.957611&z=1

If we enlarge the map and highlight the upper and lower latitudes of our Eta circle, it looks like this:

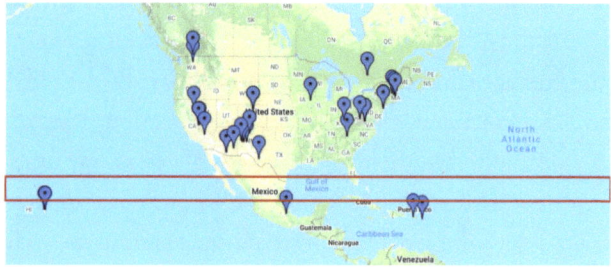

In the Western half of the world, there are no significant radio telescopes located within the Eta circle latitude boundaries. In the rest of the world, there are three located within the Eta circle latitude boundaries, shown below.

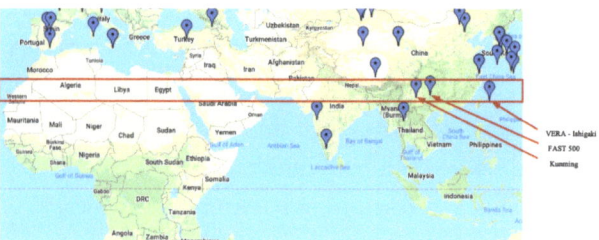

All of the money and time that is being expended for the purpose of making contact can mean only three things:

1. There are a lot of very intelligent people on Earth who believe that there must be intelligent life out there.

2. No one has established communication yet, or if someone has made contact, they are keeping it a secret.

3. We are not transmitting or listening from the right places, and/or at the right times.

Unless a station is located within the Eta circle latitude boundaries, it cannot listen or transmit in a direction perpendicular to those boundaries from Earth. Recall that we identified numerous and significant ancient sites on Earth that fall between these parallels, yet only three radio telescopes are located within the Eta circle parallels, with one of these telescopes being the newest and biggest in the world. If the goals behind constructing radio telescopes include communicating with ETs, perhaps we have selected the locations for our radio telescopes based on the wrong criteria.

The VERA Ishigakijima station is located on Ishigaki Island, which is at the southwest end of the Ryukyu Islands. VERA stands for VLBI (Very Long Baseline Interferometry) Exploration of Radio Astrometry. It also means "the truth" in Latin. It is shown below with the Milky Way visible in the night sky.

VERA is a Japanese VLBI array that is aimed at obtaining a three-dimensional map of the Milky Way galaxy. There are four systems in the VERA network.

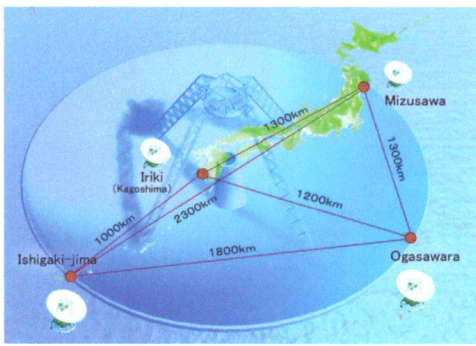

VLBI techniques measure distances and motions of radio sources in the galaxy with a very high degree of accuracy in an effort to reveal the true structure of the Milky Way galaxy. The construction of the VERA array was completed in 2002, and it has been in constant operation since 2003. The VERA station at Ishigakijima has

The world's largest and most sensitive radio telescope is the FAST 500M station...in Guizhou.

A *shisa* on the gate. This is an imaginary cross between a dog and a lion that is seen on house gates everywhere in the Ryukyu Islands. It is believed to possess a magical power that prevents evil from entering.

The world's largest and most sensitive radio telescope is the FAST 500M station located in the Dawodang depression, which is a natural basin in Pingtang County, Guizhou, southwest China. FAST stands for Five-hundred-meter Aperture Spherical Telescope. It is three times more sensitive than the Arecibo Observatory previously discussed. It is managed by NAOC/CAS (National Astronomical Observatories/ Chinese Academy of Sciences) and funded by NDRC (National Development and Reform Commission). The ultimate goal of the FAST 500M is to discover the laws of the development of the universe.

FAST 500M was officially opened for business in 2016, according to Xinhua, China's official state-run media. It did go online in 2016, but underwent a testing period through 2019. It has been given the nickname "Tianyan," which translates into "Eye of Heaven." Its first notable achievement was the discovery of two new pulsars in 2017. It is a very powerful radio telescope that is expected to make many important discoveries. The website states that its primary goals are:

1. Large-scale neutral hydrogen survey
2. Pulsar observations
3. Leading the international Very Long Baseline Interferometry (VLBI) network
4. Detection of interstellar molecules
5. Detection of interstellar communication signals
6. Pulsar timing array

In addition, the FAST 500M radio telescope will perform two sky surveys that will take about five years to complete, with an additional ten years needed just to analyze the data from those surveys. A photo of the FAST 500M is shown below; it represents an impressive scientific achievement. The FAST 500M was built as a receiver. Not much is published (that I could find) about its ability to send a message. Number 5 on the list of priorities above is what we are interested in knowing about for the purposes of this book. It is unclear how much emphasis the telescope's operators put on receiving ET communication signals or sending messages, but it is clearly not the number one priority.

Radio telescope receivers are designed to pick up radio waves from pulsars, galaxies, quasars,

FAST 500M station in Guizhou, China

and whatever else might be out there. One main obstacle for all radio telescopes is that by the time those signals reach Earth, they have decayed to nearly nothing. This is why astronomers build enormous dishes to pick up faint signals. The issue is that these dishes are sensitive to all radio waves, including interference bouncing into range from everything and anything that communicates using electricity, including engine ignition sparks and even cell phones. This is called RFI (Radio Frequency Interference). In order to minimize RFI, radio telescopes are built in remote locations such as Chile, the Australian outback, and the mountaintops of Puerto Rico.

The third radio telescope that lies within the upper and lower parallels of the Eta circle is the Kunming 40M, located in Yunnan Province on Phoenix Mountain in China. It was completed in 2006. Yunnan Observatories was originally established under the name Phoenix Mountain Observatory of Kunming at the beginning of WORLD WAR II. Since 2013, it has been operating as the Yunnan Observatories.

Yunnan Observatories

This unit was built primarily to support China's Lunar Exploration Program and its intent to launch a series of lunar probes and landers to explore the moon. The tracking system is limited by the precision needed in order to measure angular positions and the requirement of the long observation time. Its purpose is neither to listen for ET communication signals, nor to send signals out.

Of the three radio telescopes located within the Eta circle latitude boundaries, there is only one—the FAST 500M located in southwest China—that openly declares one of its objectives to be the search for communication signals from intelligent life out in the universe. If one was going to hire a giant radio telescope to test a theory about ET communication signals, the FAST 500M would be the best one. China does have something called the Telescope Access Program (TAP), which provides all China-based astronomers with access to internationally competitive medium- and large-aperture optical-infrared telescope facilities. The FAST 500M is not included as one of these facilities, so, as far as we can tell, there is no access to the facility for independent research; usage time cannot be purchased, and projects cannot be submitted for approval as with many other observatories. Then again, everything is negotiable and has a price. If the owners liked the premise and got paid enough money, they would probably work out a deal to give access for a few days.

The following diagram of how a radio telescope works is taken from: https://letstalkscience.ca/educational-resources/backgrounders/radio-astronomy.

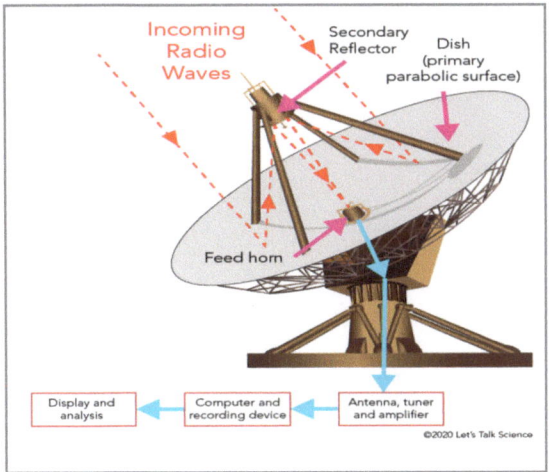

There are many similar diagrams out there. This diagram illustrates how the incoming radio waves need to hit the dish in order to be recorded. If the dish is not aimed specifically at a particular target, the desired waves will not be received. It will, however, pick up secondary RFI that drifts into its range. Our theory is that the wormhole opening is probably directly above the Eta circle. This means that our best shot at hitting that target is to get directly below it and shoot straight up. Maybe a wormhole is not something easily seen, as depicted in the fancy and colorful computer-generated interpretations of what a wormhole might look like. Maybe we have to "hear" one. It is likely that the end of a wormhole is either very silent or very noisy. It should at least be very different-sounding than the rest of the ambient noise in the area. Aiming at it will be difficult; however, we have no choice but to deduce its probable location and opening times based on the clues we have discovered.

8.
Response Implementation

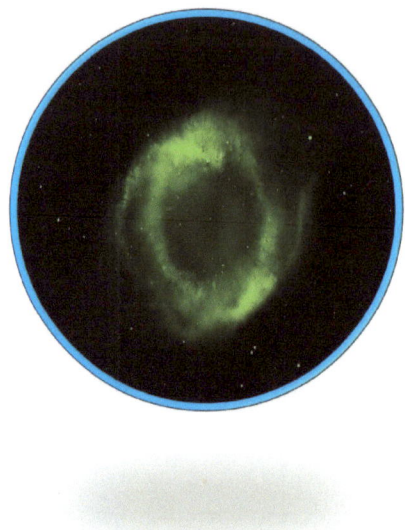

IN THEORY, WHAT COMES NEXT should be simple: send the correct response back out into the universe in the correct direction, from the correct location, and during the correct window of time. We know there is intelligent life out there, but we have failed to establish communication as far as we are aware. We have to think about what we have done in the past to establish communication, and then do our best to avoid repeating ourselves. We need to think differently about the problem and take a more simplistic approach, as outlined in this book.

What do you call someone who pounds their head against a wall over and over again because it feels so good when they stop? What do you call someone who repeats the same process over and over again, expecting a different result? If we are the kind of people who want a better result, we have to improve or completely change the process. In our case, it really does not take much to do this, so why not give it a shot?

The only way anyone can ultimately prove or disprove the ideas proposed in this book is to design a good experiment, execute the experiment, and study the results; this is the basic premise of the scientific method. It makes sense to at least try, given that what has been done historically has not achieved the desired result of establishing communication with ETs. There are various definitions of the scientific method, but the basic process is as follows.

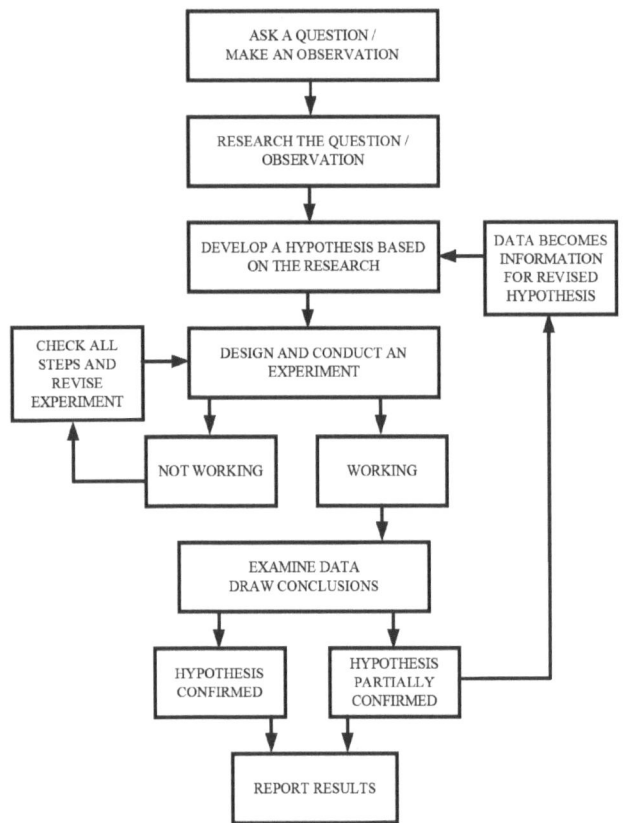

1. **Make an observation.**
   Some of the key observations made to date with regard to communication with ETs:

   a. According to eyewitness accounts and video accounts, ETs exist.

   b. Documentation exists in important ancient texts through modern times that ETs have selectively contacted individuals on Earth.

c. Ancient texts have reported that ETs have either created or influenced those cultures.

   d. We have attempted to send coded messages into the universe, but have received no documented response from ETs to those messages.

   e. Modern science believes that ETs exist and has spent billions of dollars trying to establish contact—and continues to do so.

   f. Modern science believes that wormholes exist and that their existence can be proven based on the known laws of physics; however, a wormhole has not been found to date.

   g. The observations described in this book are based on fact and lead us to a different approach to establishing ET contact as opposed to unsuccessful approaches used in the past.

2. **Ask questions about the observations and gather information.**
Are the observations listed above accurate?

   a. Yes—there are too many consistent eyewitness accounts from all over the world to conclude that they are all false.

   b. Yes—there are too many accounts throughout history from all over the world to conclude that the written accounts are all false.

   c. Yes—there is reasonable evidence to suggest that ETs have influenced human culture.

   d. Yes—to the best of our knowledge, no communication has been received.

   e. Yes—the search for communication signals from intelligent life in the universe continues at great expense and with no results.

   f. Yes—modern astrophysicists agree that wormholes must exist, but nobody has found one yet.

g. Yes—the suppositions made in this book are based on facts, and those facts lead us to the conclusion that an experiment is warranted based on a credible hypothesis.

3. **Form a hypothesis.**
A hypothesis is an explanation of the observations and the formulation of predictions based on those observations.

ET communication hypothesis:
It is possible that our attempts to initiate communication with ETs have received no response due to the following reasons:

a. We are sending the wrong messages in hopes of receiving a response. They are too complicated to decipher, and are not consistent with three-way/repeat-back communication procedures.

b. We are sending our messages in the wrong direction. We have not hit the wormhole that we need in order to close the time gap and get our messages to the area where we believe they might be received in a timely manner. We have essentially used a "message in a bottle" approach. We put our message in a vessel (satellite) and launched it with little consideration for its direction—we just hope an ET finds it.

c. We are sending our messages at the wrong times. We need to send them when a portal is open. The wormhole portal might be a predictable, but moving target.

d. The information presented in this book is sufficient to warrant an experiment in that it:

   i. Identifies the correct message format and provides a rationale.
   ii. Identifies a location from which to send the message that has a higher probability of success, again with a rationale behind it.
   iii. Identifies a time to send the message that has a higher probability of success.

e. If the correct message is sent from the correct location and at the correct time, we should receive that message (or better) back again, consistent with three-way communication procedures, in a timely manner.

4. **Test the hypothesis with an experiment that can be reproduced.**

   a. In the next section of this book, an experiment is proposed that can be reproduced.
   b. The experiment is reasonably economical compared with the investments already made that have not produced the desired results.
   c. The experiment is designed in a way that has not yet been tested. We would not be repeating a process that so far has not produced the desired result.

5. **Analyze the data and draw conclusions.**

   a. Accept or reject the hypothesis based on the data received from the experiment.
   b. Modify the hypothesis and the experiment if necessary.

6. **Reproduce/repeat the experiment with modifications as necessary.**

9.

Proposed Experiment

A REASONABLE OUTLINE FOR A repeatable experiment to test the hypothesis might look like this:

1. Research.

    a. Two different scientific organizations should create a model to find the locations in the universe where the Eta line identified in this book aligns with the orientation of Orion's belt during the times of the winter solstice and the spring equinox.

    a. Each of these organizations should conduct their own independent research to assess and verify independent results.

2. Recruit and retain a team of qualified astrophysicists who are experienced in radio telescope operations and radio signal detection. This team would be responsible for:

   a. Specifications for the equipment to be used in the experiment.

   b. Hands-on field operations and support of data collection and transmission of data.

   c. Documentation and interpretation of any radio signals that might be received during the collection of data.

3. Retain an oceanographic research vessel experienced in supporting independent scientific research. Outfit the vessel with the following:

   a. A radio telescope as specified by the astrophysics team that can transmit and receive (this includes all recording equipment for documenting any messages received).

   b. Specialized binoculars for use by qualified crews as they observe the skies throughout the duration of the experiment. These observation tools should have the capability of recording observations on video. If a target is observed, its description should be recorded as well as its bearing from the ship and azimuth. The purpose is to look for unusual movements that may indicate that the portal is open and that ETs are either coming or going.

   c. At least one optical telescope. This equipment would focus on the area(s) in the universe that the astrophysics team identified as an area where Orion is in the required alignment with our Eta Island alignment.

4. Plan an initial expedition that positions the vessel in the middle of the Eta circle (on the Tropic of Cancer) to be fully operational for three days around either the spring equinox or the winter solstice—preferably both.

5. Transmit the specified message and collect data from the radio telescope and from observations of the sky for the three specified days. It should be understood that if meaningful data is being collected, the collection of that data will not be interrupted until, in the judgment of the staff, the information has stopped being received.

6. Regardless of any information collected or not collected during the first voyage, repeat the experiment until the data covers two winter solstice events and two spring equinox events.

To the best of our knowledge, an oceangoing expedition for this purpose has never been attempted before, but I am proposing exactly that: something that has never been done before. The expedition should be privately funded to avoid any governmental interference or issues with ownership or publication of the data collected. An oceangoing radio telescope expedition would require that some unique challenges be overcome.

One challenge would be the weather, as is the case for all ships. Hurricane season has passed by the time December arrives in the Eta circle area, and it does not pick up again until May. December and March are good target dates to conduct this experiment with respect to the probability of mild weather and calmer seas. Recall that December through March is also the time frame when Orion is visible in the night sky; that fact is linked to our process. After that, Orion is on the other side of the sun. Perhaps that has something to do with the start of hurricane season in the Eta circle? That is something to think about, and a subject for another day.

USRV Atlantis: Woods Hole Oceanographic Institute

Another significant obstacle would be acquiring a suitable radio telescope and mounting it on the ship. That would be costly at face value; however, it's probably not nearly as costly as constructing, operating, and maintaining a land-based unit, and we already have plenty of those. What we do not have is an oceangoing unit that can be deployed in our target area. In order to get an expedition like this funded, a very detailed financial proposal would need to be developed. This is no different from any other expedition for whatever purpose. What is different is the value of the information that would be collected if our hypotheses are proven to be correct.

The commercial value of the data collected plus the value to humanity and even the security/defense industry value would be many times greater than the up-front cost of the investment. We would literally be opening the door to the universal community of intelligent life in the universe. We know intelligent life is out there, but so far, it has been just out of our reach.

It is our human nature to keep searching to see what is over the next hill, so we need to try new ways of getting over this hill. Perhaps what we have discovered in this book is the key to the Extraterrestrial Communication Code and real-time communication with ETs within our lifetime.

www.ingramcontent.com/pod-product-compliance
Lightning Source LLC
Chambersburg PA
CBHW040219220526

45473CB00001B/47